Sustainable Aviation

Series Editors
T. Hikmet Karakoç, Eskisehir Technical University
Eskisehir, Türkiye

C. Ozgur Colpan, Department of Mechanical Engineering
Dokuz Eylül University
Buca, Türkiye

Alper Dalkiran, School of Aviation
Süleyman Demirel University
Isparta, Türkiye

The Sustainable Aviation book series focuses on sustainability in aviation, considering all aspects of the field. The books are developed in partnership with the International Sustainable Aviation Research Society (SARES). They include contributed volumes comprising select contributions to international symposiums and conferences, monographs, and professional books focused on all aspects of sustainable aviation. The series aims at publishing state-of-the-art research and development in areas including, but not limited to:

- Green and renewable energy resources and aviation technologies
- Aircraft engine, control systems, production, storage, efficiency, and planning
- Exploring the potential of integrating renewables within airports
- Sustainable infrastructure development under a changing climate
- Training and awareness facilities with aviation sector and social levels
- Teaching and professional development in renewable energy technologies and sustainability

T. Hikmet Karakoç • Shau-Shiun Jan
Chih-Yung Wu • Currao Gaetano
Alper Dalkiran • Ali Haydar Ercan
Editors

Energy and Sustainable Aviation Fuels Solutions

Proceedings of the International Symposium on Sustainable Aviation 2023

Springer

Editors
T. Hikmet Karakoç
Faculty of Aeronautics and Astronautics
Eskisehir Technical University
Eskisehir, Eskisehir, Türkiye

Information Technology Research
and Application Center
Istanbul Ticaret University
Istanbul, Türkiye

Chih-Yung Wu
Aeronautics and Astronautics
National Cheng Kung University
Tainan City, Taiwan

Alper Dalkiran
School of Aviation
Süleyman Demirel University
Keciborlu, Isparta, Türkiye

Shau-Shiun Jan
Aeronautics and Astronautics
National Cheng Kung University
East Dist., Taiwan

Currao Gaetano
Aeronautics and Astronautics
National Cheng Kung University
Tainan City, Taiwan

Ali Haydar Ercan
UAV Technology Operator Program
Eskisehir Technical University
Eskisehir, Türkiye

ISSN 2730-7778 ISSN 2730-7786 (electronic)
Sustainable Aviation
ISBN 978-3-031-70693-6 ISBN 978-3-031-70694-3 (eBook)
https://doi.org/10.1007/978-3-031-70694-3

This Springer imprint is published by the registered company Springer Nature Switzerland AG
The registered company address is: Gewerbestrasse 11, 6330 Cham, Switzerland

Preface

The aviation sector is responsible for about two percent of global carbon emissions. With increasing concerns about climate change, the aviation industry must take significant steps to reduce its environmental impact. One of the most promising strategies for decreasing carbon emissions in aviation is the utilization of alternative jet fuels. Studies indicate that alternative jet fuels could significantly decrease carbon emissions by 50–80% compared to conventional fossil fuels. Investing in researching and deploying alternative jet fuels is an efficient strategy to achieve carbon-neutral aviation operations.

Due to increasing environmental worries and regulatory demands, there is a noticeable push toward using alternate jet fuels in the aviation sector. Global regulatory bodies require strict adherence to international carbon emission limits, prompting airlines to investigate sustainable fuel alternatives to fulfill regulatory mandates. Additionally, there is a growing domestic interest in bio-jet fuel-powered aircraft due to environmental concerns and economic benefits. The increased interest has led to expanding bio-jet fuel markets for global flights, as airlines aim to counterbalance their carbon emissions by acquiring sustainable aviation fuels and carbon offsets.

Regulated initiatives are underway in the aviation industry to promote the extensive use of alternative jet fuels and guarantee their effectiveness in commercial flights. The American Society for Testing and Materials (ASTM) has played a vital role in creating standards for bio-jet fuel, which help certify and deploy the fuel. The International Civil Aviation Organization (ICAO) establishes goals that align with the United Nations' Sustainable Development Goals. Collaborative efforts have been started to promote innovation and speed up the creation of sustainable aviation fuels. This effort includes establishing bio-jet fuel development groups and conducting test flights for fuel certification. The joint efforts highlight the industry's dedication to promoting sustainable aviation solutions and leading the shift toward a more eco-friendly and ecologically conscious future.

Although progress has been made in creating and using alternate jet fuels, obstacles remain in fully utilizing sustainable aviation technologies. Areas needing more research and development efforts are improving conversion efficiency, optimizing manufacturing processes, and lowering costs related to sustainable fuel generation. Overcoming these hurdles is crucial for promoting the widespread use of alternative jet fuels and achieving the aviation industry's goal of sustainable air travel, characterized by lower carbon emissions, improved environmental protection, and a more sustainable future for flying.

The International Symposium on Sustainable Aviation (ISSA'23), an international and multi-disciplinary symposium, addresses current issues in the research developments in the sustainable aviation field. We invited academics, scientists, engineers, practitioners, policymakers, and students to attend ISSA'23 to share knowledge, demonstrate new technologies and breakthroughs, and debate the future direction, strategies, and goals in aviation sustainability. This conference featured keynote presentations by invited speakers and general papers in oral and poster sessions.

We want to thank Springer's editorial team for their support toward the preparation of this book and the chapter authors and reviewers for their outstanding efforts.

Eskisehir, Türkiye	T. Hikmet Karakoç
Tainan, Taiwan	Shau-Shiun Jan
Tainan, Taiwan	Chih-Yung Wu
Tainan, Taiwan	Currao Gaetano
Keciborlu, Türkiye	Alper Dalkiran
Eskisehir, Türkiye	Ali Haydar Ercan

Contents

Chapter 1
Aircraft Noise Monitoring for Greening the Airports

Oleksandr Zaporozhets

Nomenclature

AEDT	Aviation Environmental Design Tool
AN	Aircraft noise
ANOMS	Airport Noise and Operations Management System
ICAO	International Civil Aviation Organization
NMT	Noise monitoring terminal

1.1 Introduction

Today, the Airport Noise and Operations Management System (ANOMS) is a sophisticated, acoustical system which monitors aircraft flight tracks, fleet mix and noise exposure levels by the location of noise monitoring terminals, time of day, season and on an annual basis. The four main ways of reporting measurements of aircraft noise were identified by the research—online platforms, reporting of noise monitor data to responsible for environmental protection governmental boards, bespoke noise reports for a given community and noise contours as an element of noise zoning around the airports. The primary goal of the ANOMS in aircraft operation is to maximize their number within environmental constraints and reduce operating costs while improving noise office productivity, reporting regulatory compliance if the noise limits are adopted and increasing community tolerance for airport growth. The last task is mostly important in the current situation—to define the way of community annoyance reduction even the traffic and appropriate noise exposure are growing. It becomes a tool necessary to be used for proactive (not simply reactive as mostly realized now around the airports worldwide) and even collaborative system of noise exposure and impact management in airports, taking

O. Zaporozhets (✉)
Department of Research Organization, Institute of Aviation (Łukasiewicz Research Network Institute of Aviation), Warsaw, Poland
e-mail: oleksandr.zaporozhets@ilot.lukasiewicz.gov.pl

T. H. Karakoç et al. (eds.), *Energy and Sustainable Aviation Fuels Solutions*, Sustainable Aviation, https://doi.org/10.1007/978-3-031-70694-3_1

in mind the International Civil Aviation Organization (ICAO) expanded the management process with including "people issues", meaning the inclusion of community engagement for aviation environmental management (Zaporozhets, 2022).

1.2 ICAO Requirements

The monitoring of aircraft noise levels in community areas is an important component of managing airport noise. Although noise monitoring has no direct noise reduction, it can be an integral component of noise management and community engagement. Aircraft noise (AN) exposure management on the population around the airports is defined by the ICAO Balanced Approach (BA). ICAO guidance on BA (ICAO Doc 9829, 2010) recommends defining the efficiency of all four fundamental elements with an assessment of day noise index Day-Night Level (DNL) at a point (area) of noise control. Overall AN exposure around the airport is usually calculated in the form of AN contours—the contour DNL = 65 dBA is defined as inappropriate (prohibited) for residential development. Requirements for AN calculations are defined by the ICAO guidance Doc 9911, noise zoning and land use—by Airport Planning Manual (ICAO Doc 9184), aircraft noise itself—by Annex 16, vol. 1 'Aircraft Noise' to ICAO Convention (ICAO Annex 16, vol. 1, 2019).

The Airport Planning Manual (ICAO Doc 9184, 2018) in Appendix 5 describes monitoring as 'to be the routine measurement of noise levels created by aircraft in the operation of an aerodrome. Monitoring usually involves a large number of measurements per day, from which an immediate indication of the noise level may be required". In wider content AN monitoring reports, which include radar flight tracks, modelled or measured noise, and weather data, can provide an objective source of information for all stakeholders and local communities close to the airport, to assess the effectiveness of noise management procedures around the airport (Fig. 1.1). The quality of noise exposure assessment is dependent on Noise monitoring terminal (NMT) number and their location around the airport—how their measured values are a representation for exposure assessment for the traffic under consideration. Coordination and collaboration between the stakeholders (between the airport authority and local community first of all) at the level of ANOMS data management is usually absent, including the possible interdependences and trade-offs between environmental factors (emission/pollution in atmosphere and fuel burn first of all). ICAO policy in protection from AN impact is defined as a reduced number of people affected by noise—in most cases, the people are annoyed because of the number of possible noise disturbances. If doing nothing with annoyance management, the future scenarios for air transportation are looking unsatisfactorily, because existing results of the studies of noise annoyance in other sectors, different from aviation and having the same or quite similar acoustical properties, for example, wind farm noise, show the much higher annoyance than existing aircraft noise studies.

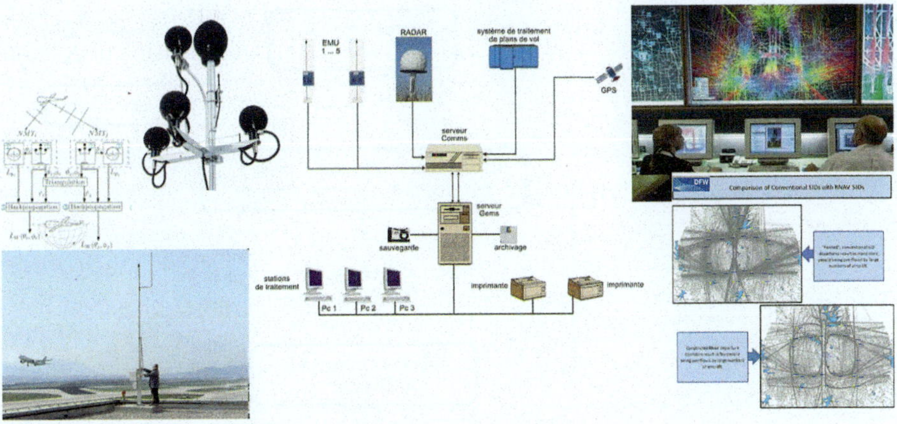

Fig. 1.1 Noise exposure measurement approach for current ANOMS in airports

The place of the ANOMS in the overall noise management program of the airports is not defined by ICAO still. Every airport authority (in the better case with the influence of other stakeholders) defines the contribution to their AN protection programs from its point of view and experience only. The technical requirements from the standards (SAE ARP4721, 2012; ISO 20906, 2009) are not good enough for the successful operation and maintenance of the ANOMS. A specific ICAO document must specify the definition of ANOMS and their place in the noise management program at airports, reasons for continuous or portable noise measurement in airports and the goal and purposes for NMT installations in airports.

1.3 Methodology for Analysing and Reporting Noise Monitor Data

Like the procedure of handling noise complaints by current ANOMS, it should be ensured that the reporting on noise monitor data is understandable to local communities and is focussed on identifying potential issues and tangible solutions. First of all, it means that ANOMS must provide continuing measurements at their separated NMT—the noise levels and acoustical samples during very short intervals (sometime around 20 ms) during the whole day/week/month/year. The second point is a contemporary requirement to identify the source of noise—maybe an AN over the background noise, in a better case in smart autonomous mode. Signal classification must be implemented as the process of identifying the object associated with a given input signal and once the features have been extracted as described, an n-dimensional feature vector will input a classifier. The third important point for understanding the exposure by aircraft noise source, at least in what direction to NMT the source is located (if a moving source—moving/flight path design should be provided). This included agreeing with communities what outputs and metrics

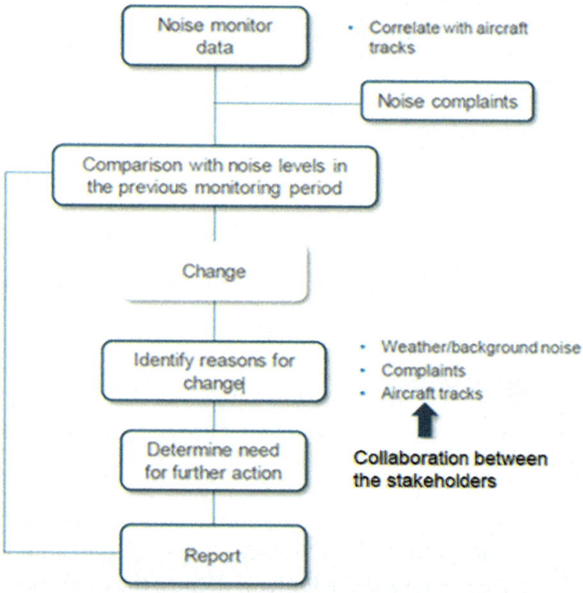

Fig. 1.2 Research methodology for the noise assessment of the novel aircraft

Fig. 1.3 WebTrak system showing a noise event (in red) as registered by a noise monitor

are meaningful to them and a standard methodology for analysing noise monitor data (Figs. 1.2 and 1.3).

The study in 2009 (Hooper et al., 2010) showed that current indicators, when used in regulations and for community explanations in the same way, have the serious drawback of being too technical for non-experts and not meeting the expectations of citizens. Residents do not understand these metrics, nor do they rely on them. Today, some airports use complementary indicators that give information that can be more easily understood by a community like "Number Above", "Time Above", "Person-Events Index" and "Average Individual Exposure". The challenge of further ANOMS improvements is to present this data in a way that is understandable to communities, leads to the identification of any issues and, as appropriate, the need for further action to manage the problem with benefit for all stakeholders. The airports should undertake the review of locations for temporary NMT to support the understanding of issues.

Standards for noise communication tools to the public, similar with technical requirements (SAE ARP4721, 2012; ISO 20906, 2009), are absent. A variety of indicators are used to show the noise exposure value or a part of it that may dismiss (preclude) the correct understanding by people and following reaction to noise exposure. A specific standard (a new ICAO document or international standard) must specify the noise data, their metrics and their indication for non-expert people—noise communication to the public must be unified. Also, the importance of AN management evolution in airports to proactive and collaborative levels requires specific abilities of the ANOMS: to predict the noise exposure (not only to measure and assess the current levels) in line with the vision of airport and especially airlines' visions of the development; to define the compliance with requirements and new conditions (even the air transportation is changed by introduction of air taxies and unmanned vehicles in traffic to be closer to resident demands); to communicate effectively with non-experts and efficiently cooperate with them in AN management (not only for the specific tasks like AN level control at sensitive locations inside the community); to be transparent for the community as a specific condition for cooperation with it, etc.

1.4 Conclusion

ICAO efforts should be provided in preparation for new guidelines for airports and communities to describe the goal and purposes of the ANOMS in the overall collaborative noise management program at the airports, best practice analysis worldwide, its functional capabilities in noise measurements and data processing, reporting and visualization, including Web applications for community information and involvement in noise managing activities by including a special chapter on how to reduce an impact on the communities with similar exposure of aircraft noise.

Besides the technical elements, which are based on AN exposure measurements and appropriate metrics used, the noise annoyance as the most essential kind of outcomes of AN exposure to neighbouring communities must now be addressed. This evolution may also lead to a new vision of the ICAO balanced approach to aircraft noise control in very near future with quite fast introduction of urban air mobility, which should complement existing air transportation system and being much closer to the residents in vicinity of usual airports.

References

ICAO Annex 16. (2019). *Annex 16 to the convention on international civil aviation Vol. I – Aircraft Noise*. ICAO. https://store.icao.int/en/annex-16-environmental-protection-volume-i-aircraft-noise

ICAO Doc 9184 – Part 2. (2018). *Airport planning manual – Part II – Land use and environmental management* (4th ed.). https://store.icao.int/en/airport-planning-manual-land-use-and-environmental-management-doc-9184-part-2

ICAO Doc 9829. (2010). *Guidance on the balanced approach to aircraft noise management*. ICAO Doc 9829 (2nd ed.). ICAO. https://store.icao.int/en/guidance-on-the-balanced-approach-to-aircraft-noise-management-doc-9829

ISO 20906. (2009). *Acoustics. Unattended monitoring of aircraft sound in the vicinity of airports*. https://www.iso.org/standard/35580.html

Hooper, P. D., Thomas, C. S., Maughan, J. A., Hume, K. I (2010). Aircraft Noise and Community Impacts. In book: Encyclopedia of Aerospace Engineering. https://doi.org/10.1002/9780470686652.eae343

SAE ARP4721. (2012). *Monitoring aircraft noise and operations in the vicinity of airports: system description, acquisition, and operation*. SAE. https://www.sae.org/standards/content/arp4721/1

Zaporozhets, O. (2022). Balanced approach to aircraft noise management. In *Aviation noise impact management. Technologies, regulations, and societal well-being in Europe*. https://link.springer.com/chapter/10.1007/978-3-030-91194-2_3

Chapter 2
Ranking of Indian Airlines

Mayank and T. Bangar Raju

Nomenclature

DEA	Data envelopment analysis
DMUs	Decision-making units
OTP	On time performance

2.1 Introduction

Data envelopment analysis (DEA) is a nonparametric approach used to evaluate the relative efficiency of a set of decision-making units (DMUs) based on their inputs and outputs. It was developed in the 1970s by Charnes, Cooper, and Rhodes and has since become a widely used technique in operations research, management science, and other fields (Wang et al., 2016). The primary aim of DEA is to identify the most efficient DMUs among a set of entities that share similar goals but may use different combinations of inputs and outputs to achieve them (Pannala et al., 2022). This technique assumes that each DMU can be characterized by a set of inputs and outputs, and that the inputs are transformed into outputs using a production function. DEA calculates an efficiency score for each DMU by comparing its performance with that of the other DMUs in the set and identifying the best-performing units as the most efficient (Rahim et al., 2018). DEA has been used in a wide variety of applications, including evaluating the efficiency of banks, hospitals, schools, and other public services. It is particularly useful in cases where traditional statistical methods are not applicable, such as when the data are too complex or too scarce, or when the underlying production process is unknown or cannot be easily modeled (Hafezalkotob & Hafezalkotob, 2015). A DMU is an entity that produces outputs from inputs, such as a firm, a hospital, a school, or a government agency. DEA

Mayank (✉) · T. B. Raju
School Of Business, University of Petroleum and Energy Studies, Dehradun, India

T. H. Karakoç et al. (eds.), *Energy and Sustainable Aviation Fuels Solutions*, Sustainable Aviation, https://doi.org/10.1007/978-3-031-70694-3_2

allows us to compare the efficiency of different DMUs by analyzing how much output each unit can generate relative to its inputs while taking into account the different combinations of inputs used by each unit. The goal of DEA is to identify the DMUs that are operating at the frontier of efficiency, meaning that they are using their inputs to generate the maximum amount of output possible (Chauhan et al., 2017). In this chapter, I will be comparing five Indian airlines based on various DMUs, and using those DMUs we will be able to compare and find the most efficient airline in terms of operational performance in the Indian market. I will use data from the years 2016–2019. During this period, markets were at a peak and later got affected due to the pandemic and every other airline suffered because of it. Data from 2016 to 2019 will tell us about the peak performance of various airlines and how they managed to grow over time with it.

2.2 Research Methodology

We have used a nonparametric research strategy in my study. We have performed a DEA analysis with the secondary data collected from five largest airlines based in India. This study relies on secondary data. All the data has been collected from annual reports of airlines and Directorate General of Civil Aviation (DGCA) and Ministry of Civil Aviation (MoCA) websites. Techniques that are used for the study are DEA and Malmquist productivity index.

2.2.1 *Decision-Making Units (DMUs)*

We will be using the following DMUs for comparison of various airlines:

- Passengers flown.
- Market share.
- On-time performance (OTP).
- Air connectivity.
- Fleet count.
- Flights operated.

2.3 Analysis and Findings

The Malmquist Index Summary of Firm Means provides a comprehensive evaluation of the productivity of airlines based on their total factor productivity change over a certain period (Table 2.1). The results show that none of the airlines in the analysis has demonstrated a positive productivity trend. This means that each

Table 2.1 Indian airlines productivity efficiency results

Malmquist index summary of firm means					
Firm	Efficiency change	Technical efficiency change	Pure efficiency change	Scale efficiency change	Total factor productivity change
Indigo	1.000	0.955	1.000	1.000	0.955
Spice Jet	0.973	1.004	0.987	0.985	0.977
Air India	0.871	1.041	0.882	0.987	0.906
GoAir	1.000	0.973	1.000	1.000	0.973
Air Asia	1.000	0.815	1.000	1.000	0.815
Mean	0.967	0.954	0.973	0.994	0.923

airline's output relative to their input has either decreased or remained stagnant. However, there are some differences in the productivity trends among the airlines. Spice Jet had the highest total factor productivity change of all airlines analyzed, with a score of 0.977, which is close to 1. This indicates that Spice Jet's efficiency has improved significantly compared to other airlines, as it has achieved a near-ideal balance between its inputs and outputs. Following Spice Jet, GoAir, and Indigo had relatively better total factor productivity changes, with scores of 0.973 and 0.955, respectively. While these scores are not as high as Spice Jet's score, they still suggest that GoAir and Indigo have also made some progress in improving their productivity. Overall, the Malmquist Index Summary of Firm Means provides valuable insights into the performance of airlines and highlights areas where they can focus to enhance their productivity. By identifying areas for improvement, airlines can develop strategies to optimize their inputs and outputs, enhance their efficiency, and maintain a competitive edge in the market.

2.4 Conclusion

This chapter uses Malmquist DEA to measure productivity and efficiency improvement among five Indian airlines; these are the airlines with the highest market presence. We observe that Spice Jet airline has the maximum total factor productivity change while Air Asia is the lowest. The advantage of Spice Jet airlines during this period of study is that they operated very less flights as compared to the market leader Indigo. They could focus more on OTP and other factors which resulted in better operational performance of the airlines. Meanwhile, Indigo despite being the market leader and operating a lot of flights and best connectivity among other airlines could not maintain their operational parameters and hence was ranked third in the analysis. It really does not depend on the fleet size and the number of passengers that the airline carries; they need to have control over every parameter that contributes to operational efficiency. Few airlines like Spice Jet, Air India, and GoAir had an MPI score more than one during individual years but failed to have a score greater than one in the firm's annual mean. During the period of 2016–2019, none of the airlines had an MPI score higher than one in the firm's annual mean, but the values came out to be close to one; hence, the rankings are based on the resultant values.

Following are the rankings for the airlines using the study:

- Spice Jet Airlines.
- GoAir.
- Indigo Airlines.

References

Chauhan, R., Singh, T., Tiwari, A., Patnaik, A., & Thakur, N. S. (2017). Hybrid entropy – Topsis approach for energy performance prioritization in a rectangular channel employing impinging Air Jets. *Energy, 134*, 360–368. https://doi.org/10.1016/j.energy.2017.06.021

Hafezalkotob, A., & Hafezalkotob, A. (2015). Extended MULTIMOORA method based on Shannon entropy weight for materials selection. *Journal of Industrial Engineering International, 12*(1), 1–13. https://doi.org/10.1007/s40092-015-0123-9

Pannala, R. K. P., Mogha, S. K., Sharma, N., & Garg, V. (2022). Combined Shannon's entropy and DEA: Case of Indian hotel and restaurant sector. *International Journal of Data Analysis Techniques and Strategies, 14*(3), 203.

Rahim, R., Supiyandi, S., Siahaan, A. P., Listyorini, T., Utomo, A. P., Triyanto, W. A., Irawan, Y., Aisyah, S., Khairani, M., Sundari, S., & Khairunnisa, K. (2018). Topsis method application for decision support system in Internal Control for selecting best employees. *Journal of Physics: Conference Series, 1028*, 012052. https://doi.org/10.1088/1742-6596/1028/1/012052

Wang, C.-N., Lin, H.-S., Hsu, H.-P., Le, V.-T., & Lin, T.-F. (2016). Applying data envelopment analysis and grey model for the productivity evaluation of Vietnamese agroforestry industry. *Sustainability, 8*(11), 1139. https://doi.org/10.3390/su8111139

Chapter 3
A Brief Assessment of Aircraft Fuel Consumption and Pollutant Emissions for Departure Operations

Kadir Dönmez and Ramazan Kursat Cecen

Nomenclature

ADS-B	Automatic dependent surveillance-broadcast
HC	Hydrocarbons
CO	Carbon monoxide
NOx	Nitrogen oxides
LTO	Landing and take-off
ICAO	International Civil Aviation Organization

3.1 Introduction

The aviation sector has been a major force in driving global trade, communication, business, and tourism. Nevertheless, the industry still confronts several challenges despite its rapid development. These include increasing fuel consumption, rising fuel costs, growing air traffic, increased competition, economic crises, aviation pollutants, safety issues, and design and operational challenges. Despite these difficulties, the aviation industry continues to play a critical role in fostering global connectivity and economic progress. One of the greatest difficulties facing the aviation sector is fuel use. With rising oil prices and increased competition, fuel consumption has evolved into a critical issue that has been demanding attention for a considerable period. As a result of burning fuel, aircraft engines emit pollutants.

K. Dönmez (✉)
Aircraft Maintenance Department, Samsun University, Samsun, Turkiye
e-mail: kadir.donmez@samsun.edu.tr

R. K. Cecen
Motor Vehicles And Transportation Technology Department, Eskisehir Osmangazi University, Eskişehir, Turkiye
e-mail: ramazankursat.cecen@ogu.edu.tr

T. H. Karakoç et al. (eds.), *Energy and Sustainable Aviation Fuels Solutions*, Sustainable Aviation, https://doi.org/10.1007/978-3-031-70694-3_3

The environmental concerns arising from these emissions are significant because of their global impact and effect on local air quality at ground level (Lee et al., 2010; Larsson et al., 2019; Brzozowski & Kotlarz, 2005). Therefore, it is vital to address these concerns and identify measures to reduce the impact of aircraft emissions. Pollutant emission inventories are essential for assessing pollution, simulating the environment, and developing pollution control strategies. An entire flying mission can be separated into two parts: the landing and takeoff (LTO) cycle and the cruising phase. Consequently, analyzing aircraft emissions entails monitoring airport ground-level pollutant emissions and evaluating emissions during the cruise phase (Fan et al., 2012).

According to the International Civil Aviation Organization (ICAO), the LTO cycle consists of four flight phases: taxi/idle, approach, climb-out, and takeoff. In this study, we focused only on the ground operations. The proposed study uses the ICAO Engine Emissions Databank to calculate the fuel burn during ground operations. This databank provides fuel burn indices for each flight phase and assumes specific thrust settings for each ground operational phase: 7% for taxi/idle, 30% for approach, 85% for climb-out, and 100% for take-off (Kim & Baik, 2020). In this study, we collect the ADS-B data using FlightRadar24 and determine the taxi time for each aircraft. Then, aircraft engine types and their parameters are obtained using the ICAO Engine Emissions Databank (ICAO, 2023). Moreover, the fuel consumption and pollutant emissions values are calculated for each aircraft.

Numerous studies have been conducted to calculate air pollutant emissions from aircraft during landing and takeoff (LTO) cycles, including nitrogen oxides (NOx), carbon monoxide (CO), and hydrocarbons (HC). An early examination at Tbilisi International Airport was undertaken by Tokuslu (2020), who gave crucial details about how aviation affects local air quality. Similar studies conducted at Kayseri Airport by Ylmaz (2017) and at Adnan Menderes Airport by Tolga (2008) have emphasized the significance of comprehending the environmental impact of the aviation industry for a safer and cleaner world. Additionally, Kim and Baik have suggested creative ways to use airplane trajectory data to estimate fuel burn during ground operations (2020). Their research focuses on how taxi phases and related thrust settings affect fuel consumption. Based on flight data recorder (FDR) data, Khadilkar and Balakrishnan (2012) suggested a fuel prediction model, which identified acceleration events and taxi duration as key influences on fuel use during taxiing. Collectively, these studies highlight how essential emission studies are to developing aviation practices that are environmentally friendly.

To contribute to the previous efforts, we evaluated the departure fuel consumption and emissions at Istanbul Sabiha Gökçen International Airport considering the aircraft ground trajectories and engine types based on ADS-B and ICAO database, respectively. Then the obtained results were compared to the results of the previously reported model by Khadilkar and Balakrishnan (2012).

3.2 Method

Aircraft taxi durations were estimated using the ADS-B data set, and the operation took place on June 4, 2023, at Istanbul Sabiha Gökçen International Airport. The departure operations were conducted between 07:00 and 08:00 for 18 different aircraft. The ADS-B data cover timestamps, UTC, aircraft type, callsign, position, altitude, speed, and heading information. The taxi duration for an aircraft starts right after the pushback operation and finishes right before the take-off roll. When estimating fuel burn for an aircraft, it is essential to factor in the taxiing phase and the type of engine being used. Fuel flow rate indices specific to each engine type must be considered during the fuel-burn calculation. Considering these factors, three different aircraft types and engine types were considered, and their characteristic is presented in Table 3.1.

3.3 Results

In this section, the fuel consumption and pollutant emissions values are presented. Each aircraft taxi duration is derived from the ADS-B data set and the fuel flow rate for idle thrust configuration for each engine type is determined. According to the taxi duration, fuel consumption values and pollutant emissions values are calculated and provided in Table 3.2.

According to the results, average fuel consumption values are determined as 234.95 kg. Also, maximum fuel consumption and minimum fuel consumption are found as 392.99 kg and 106.37 kg, respectively, for the departure aircraft set. The total pollutant emissions value is 145.3 kg, and their percentage distribution is presented in Fig. 3.1.

Furthermore, average CO, NOx, and HC emission values are determined as 6.04 kg, 1.23 kg, and 0.81 kg per aircraft, respectively. In addition, maximum and minimum values are obtained as 13.95 and 1.59 kg for CO emission, while these values correspond to 2.17 kg and 0.51 kg for NOx emission. Then, maximum and minimum values are obtained as 2.17 kg and 0.18 kg for HC emission.

Table 3.1 Characteristic aircraft and engine arrangements for the entire calculation

Aircraft type	Engine identification	Fuel flow idle (kg/sec)	HC EI idle (g/kg)	CO EI idle (g/kg)	NO_x EI idle (g/kg)
Airbus A320	CFM56-5B6	0.13	2.7	35.5	4.2
Airbus A321	CFM56-5B2/2P	0.111	3.4	46.1	3.9
Boeing 737-8	CFM56-7B27	0.116	1.7	17.9	4.8

Table 3.2 Aircraft fuel consumption and pollutant values

Aircraft number	Aircraft type	Taxi duration (Sec)	Fuel (kg)	CO EI idle (kg)	NO_x EI idle (kg)	HC EI idle (kg)
1	Airbus A320	1732	225.16	7.99	0.95	0.61
2	Airbus A320	1947	253.11	8.99	1.06	0.68
3	Airbus A320	1868	242.84	8.62	1.02	0.66
4	Airbus A320	2004	260.52	9.25	1.09	0.70
5	Airbus A320	2903	377.39	13.40	1.59	1.02
6	Airbus A320	3023	392.99	13.95	1.65	1.06
7	Airbus A320	2021	262.73	9.33	1.10	0.71
8	Airbus A321	2177	241.65	2.17	2.17	2.17
9	Airbus A321	1813	201.24	1.81	1.81	1.81
10	Airbus A321	1592	176.71	1.59	1.59	1.59
11	Airbus A321	960	106.56	4.91	0.96	0.96
12	Boeing 737-8	917	106.37	1.90	0.51	0.18
13	Boeing 737-8	2313	268.31	4.80	1.29	0.46
14	Boeing 737-8	1398	162.17	2.90	0.78	0.28
15	Boeing 737-8	2182	253.11	4.53	1.21	0.43
16	Boeing 737-8	1729	200.56	3.59	0.96	0.34
17	Boeing 737-8	2618	303.69	5.44	1.46	0.52
18	Boeing 737-8	1672	193.95	3.47	0.93	0.33
Total		1732	4229.07	108.65	22.14	14.51

3.3.1 Comparison of Results

The fuel consumption values are obtained based on the ICAO emission databank, and the results of the regression model presented by Khadilkar and Balakrishnan (2012), are compared in this section. They presented a linear regression model based on flight data recorder (FDR) data, given in Eq. (3.1), considering the number

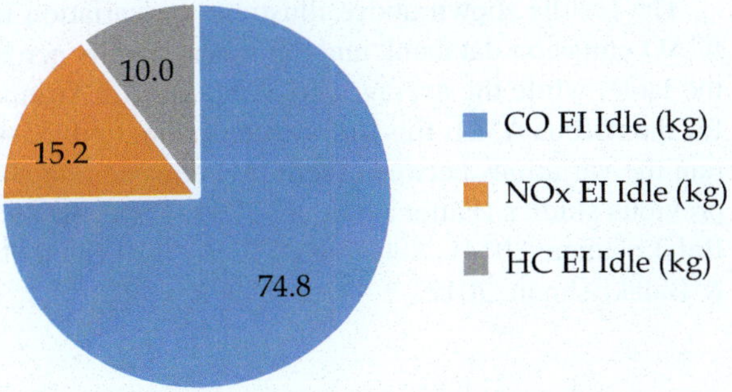

Fig. 3.1 Pollutant emissions distribution

Table 3.3 Comparison of ICAO databank and regression model results

Aircraft number	Aircraft type	Taxi time (sec.)	Number of ACC events	Fuel consumption per aircraft (kg)		Difference (%)
				ICAO databank	LRM	
1	A320	1732	3	450.32	368.93	18.07
2	A320	1947	1	506.22	410.19	18.97
3	A320	1868	2	485.68	395.56	18.56
4	A320	2004	2	521.04	424.18	18.59
5	A320	2903	2	754.78	613.36	18.74
6	A320	3023	1	785.98	636.62	19.00
7	A320	2021	2	525.46	427.75	18.59
8	A321	2177	3	483.29	482.42	0.18
9	A321	1813	2	402.49	401.32	0.29
10	A321	1592	2	353.42	352.94	0.14
11	A321	960	2	213.12	214.59	−0.69

of acceleration events and taxi times of departures as variables. They defined the acceleration event as the aircraft accelerating at more than 0.15 m/s^2 for at least 10 s. Despite the inherent difficulties in determining acceleration events from ADS-B data, it was feasible to manually compute them for our samples due to the small sample size. However, the presence of time jumps in the ADS-B data poses several challenges when it comes to accurately calculating acceleration events especially higher number of samples.

$$\frac{f}{\sqrt{T_{\text{amb}}}} = a1 + b1 \cdot t + c1 \cdot n_a \tag{3.1}$$

In the given Eq. (3.1), 'f' denotes the fuel consumed, 't' signifies the taxi time, and 'n_a' represents the number of acceleration events made by aircraft during taxi. The parameters to be estimated are '$a2$', '$b1$', and '$c1$'. The comparison of fuel calculated based on the ICAO emission databank and the regression model is given in Table 3.3.

The results shown above illustrate the variation in the differences between the ICAO emission databank and the model results across aircraft types. According to the table, while the average difference for the Airbus A320 is approximately 18%, for the Airbus A321, this difference is a negligible –0.02%. These results corroborate the variations between the ICAO emission databank and FDR results found in previous studies (Patterson et al., 2009: Khadilkar & Balakrishnan, 2012; Chati & Balakrishnan, 2014). Note that B737 has not been included in the study of Khadilkar & Balakrishnan, 2012.

3.4 Discussion and Conclusion

Our research evaluates the levels of aviation fuel consumption and pollutant emissions during taxi operations at Istanbul Sabiha Gokçen Airport. We have utilized the ADS-B data to filter solely on taxi operations, followed by an analysis of aircraft engine types. By determining their respective parameters using the ICAO engine emissions databank, we can calculate fuel consumption, as well as HC, CO, and NOx. All calculations are based on the idle configuration provided in the ICAO database. However, during taxi movements, while aircraft are on the ground, fuel consumption and emission values can vary due to frequent stops and starts, turns, or acceleration events. Departures may experience these events since the arrivals have priorities during the operations (Khadilkar & Balakrishnan, 2012). Considering these variations in calculations allows for more precise results to be obtained. Although the study proposed by Nikoleris et al. (2011), provide a detailed estimate of fuel consumption and emissions during taxi operations at Dallas/Fort Worth International Airport, suggest that stop-and-go situations, primarily due to taxiway congestion, account for approximately 18% of fuel consumption, Khadilkar and Balakrishnan's (2012) research provides a different viewpoint. According to the 2012 study by Khadilkar and Balakrishnan, taxi-out time and acceleration events are more significant predictors of taxi-out fuel use than the number of stops and turns. Due to the modest impact of these variables and their varying significance between aircraft types, their elimination may improve the predicted accuracy of the model. In addition, the operational complexity of tracking stops and turns strengthens their argument. However, applying this model to current research was difficult because of the time jumps in the ADS-B data obtained from flightradar24, which renders the calculation of acceleration events difficult. Consequently, while the findings of Khadilkar and Balakrishnan provide important insights, their applicability may be limited by data restrictions and research settings. These further mandates the use of ICAO emission databank settings or machine learning methods to correct for the missing ADS-B data, especially in light of the limited information available in comparison to FDR data. Overall results indicate that there is significant variation in the differences between the ICAO emission databank and the model results across aircraft types. This approach, however, may provide only approximate calculations for fuel and emission data when access to FDR data is unavailable.

References

Brzozowski, K., & Kotlarz, W. (2005). Modelling of air pollution on a military airfield. *Atmospheric Environment, 39*(33), 6130–6139. https://doi.org/10.1016/j.atmosenv.2005.06.040

Chati, S. Y., & Balakrishnan, H. (2014). *Analysis of aircraft fuel burn and emissions in the landing and take off cycle using operational data*. In 6th International Conferenceon Research in Air Transportation (ICRAT2014).

Elbir, T. (2008). Estimation of engine emissions from commercial aircraft at a midsized Turkish airport. *Journal of Environmental Engineering, 134*(3), 210–215. https://doi.org/10.1061/(ASCE)0733-9372(2008)134:3(210)

Fan, W., Sun, Y., Zhu, T., & Wen, Y. (2012). Emissions of HC, CO, NOx, CO2, and SO2 from civil aviation in China in 2010. *Atmospheric Environment, 56*, 52–57. ISSN 1352-2310. https://doi.org/10.1016/j.atmosenv.2012.03.052

ICAO. (2023). *Aircraft engine emissions databank*. European Aviation Safety Agency.

Khadilkar, H., & Balakrishnan, H. (2012). Estimation of aircraft taxi fuel burn using flight data recorder archives. *Transportation Research Part D: Transport and Environment, 17*(7), 532–537, ISSN 1361-9209. https://doi.org/10.1016/j.trd.2012.06.005

Kim, H. J., & Baik, H. (2020). Empirical method for estimating aircraft fuel consumption in ground operations. *Transportation Research Record, 2674*(12), 385–394. https://doi.org/10.1177/0361198120961033

Larsson, J., Elofsson, A., Sterner, T., & Åkerman, J. (2019). International and national climate policies for aviation: A review. *Climate Policy, 19*(6), 787–799. https://doi.org/10.1080/14693062.2018.1562871

Lee, D. S., Pitari, G., Grewe, V., Gierens, K., Penner, J. E., Petzold, A., Prather, M. J., Schumann, U., Bais, A., Berntsen, T., Iachetti, D., Lim, L. L., & Sausen, R. (2010). Transport impacts on atmosphere and climate: Aviation. *Atmospheric Environment, 44*(37), 4678–4734. https://doi.org/10.1016/j.atmosenv.2009.06.005

Nikoleris, T., Gupta, G., & Kistler, M. (2011). Detailed estimation of fuel consumption and emissions during aircraft taxi operations at Dallas/Fort Worth international airport. *Transportation Research Part D: Transport and Environment, 16*(4), 302–308. ISSN 1361-9209. https://doi.org/10.1016/j.trd.2011.01.007

Patterson, J., Noel, G. J., Senzig, D. A., Roof, C. J., & Fleming, G. G. (2009). Analysis of departure and arrival profiles using real-time aircraft data. *Journal of Aircraft, 46*(4), 1094–1103. https://doi.org/10.2514/1.42432

Tokuslu, A. (2020). Estimation of aircraft emissions at Georgian international airport. *Energy, 206*, 118219. https://doi.org/10.1016/j.energy.2020.118219

Yılmaz, İ. (2017). Emissions from passenger aircraft at Kayseri Airport, Turkey. *Journal of Air Transport Management, 58*, 176–182. https://doi.org/10.1016/j.jairtraman.2016.11.001

Chapter 4
Shear Thickening Fluid–Based Triboelectric Nanogenerators

Selim Gürgen

4.1 Introduction

Smart wearable electronic systems containing signal perception and recognition, data processing, data transmission, reception, and reprocessing show wide applications in e-skin, human-machine interaction, and smart robots. Precisely detecting and sensing various external stimuli was the chief premise for the normal operation of these systems. Particularly, based on triboelectrification and electrostatic induction, triboelectric nanogenerator (TENG), as a self-powered sensor device with high sensitivity, has shown favorable advantages in smart electronics due to its wide material availability, high efficiency, and flexible usage. Based on the triboelectric effect and electrostatic induction, TENG can convert various kinetic energy into electricity. Especially, with the increasing use of advanced soft electronics, wearable TENGs, owing to their advantages of high flexibility and efficiency, could be attached to the human body showing important potential for application as portable power sources (Yun et al., 2022).

To further improve wearability and comfortability, various fabric-based TENGs have been designed so far. Textile, owing to the advantages of softness, low price, and wearing comfort, is an ideal candidate for designing wearable TENG devices. Additionally, elastic multifunctional fiber with the energy-harvesting property from body motion was presented which could be served as self-powered tactile sensors for wireless music controllers and human-machine interfaces. Nevertheless, traditional fiber-based TENGs with weak mechanical performance are easily damaged under harsh impact loadings, resulting in structure destruction and electrical performance degradation. Consequently, it is necessary to develop novel fiber-based

S. Gürgen (✉)
Department of Aeronautical Engineering, Eskişehir Osmangazi University, Eskişehir, Türkiye
e-mail: sgurgen@ogu.edu.tr

T. H. Karakoç et al. (eds.), *Energy and Sustainable Aviation Fuels Solutions*, Sustainable Aviation, https://doi.org/10.1007/978-3-031-70694-3_4

TENG with the enhanced anti-impact performance which could be used for intelligent protective clothing (Wang et al., 2020).

Recently, there was a successful strategy to improve the mechanical properties of high-performance textiles such as Kevlar and UHMWPE by introducing polymer composites on the fabric. Especially, it has been proven that shear-thickening materials with rate-dependent mechanical properties could significantly enhance the fiber frictions as well as the anti-impact performance of the composites. Shear thickening materials contain shear thickening fluid (STF) and shear stiffening gel (SSG). The viscosity of non-Newtonian STF can dramatically increase when the shear rate exceeds a critical value and recover immediately after unloading the applied shear rate. STF can dissipate impact energy, and it has been extensively applied in various areas including anti-impact and damping.

Because the current wearable TENGs were invalid under impact conditions, high-speed kinetic energy could not be harvested by them. They also could not detect consecutive impact excitation as self-powered sensors. Some functional TENGs with excellent self-healing properties could recover their triboelectric as well as mechanical properties after destruction, which surely prolonged their lifespan. However, these polymer-based TENGs still could not maintain a violent impact and harvest the energy. Thus, it may be possible to develop all-in-one functional TENGs with good protection and the ability to collect various kinds of dynamic impact energies by combining STF/SSG with Kevlar fibers.

4.2 Literature Survey

There are many research reports in the literature that focused on improving the impact resistance performance of high-performance textiles with STF and SSG. For instance, an intelligent protective textile based on graphene/Kevlar showed self-powered monitoring of physiological movements and hazardous gases properties. Besides good thermal management and self-powered sensing properties, a personal wearable Kevlar device could also resist low-velocity impact loadings. Recently, there was a successful strategy to improve the mechanical properties of Kevlar by introducing polymer composites to the fabric. Especially, it has been proven that shear-thickening materials with rate-dependent mechanical properties could significantly enhance the fiber frictions as well as the anti-impact performance of the composites. It was found that the yarn pull-out force of Kevlar fabrics treated with graphene oxide and carbon nanotubes multi-phase STF was almost two times that of neat Kevlar. Additionally, when Kevlar fabric contains 34.89 wt.% STF (STKF), the stab resistance of the eight layers of STKF was enhanced remarkably than that of ten layers of neat Kevlar fabric. Moreover, SSG, a derivative of polyborosiloxane, is a soft polymer, which shows a shear-stiffening effect similar to STF. Its storage modulus increases 3 to 4 orders of magnitude with the increase of shear frequency. Importantly, the solid SSG is favorable to protecting STF from moisture and heat. By combining Kevlar with shear-stiffening polymers, the penetration impact energy of the composite represented a 50% increment with neat Kevlar (Wang et al., 2022).

On the other hand, it was reported that a portable TENG made of nylon and Dacron fabric could effectively harvest body motion energy and generate a voltage as high as 2 kV. Smart TENGs relying on cotton socks could capture and reveal more direct sensory information on the human body including physiological motions, contact force, and sweat levels. The introduction of polyacrylonitrile into TENG endowed the device with machine-washable and breathable properties. Besides, as miniaturization and integration are more favorable for wearable devices, various all-in-one textile-based TENGs have been further developed by taking advantage of the superiority of fibers. Besides harvesting energy, the polyester fiber-based TENG also showed good disinfection performance toward bacteria. Because of the flame-retardant yarn, a 3D honeycomb-structured TENG could work as a self-powered escape in fire rescue. These results indicate functional textiles will endow TENGs with additional properties that widen their practical applications. However, currently reported TENG devices only harvest slight mechanical energy, which includes human motion, water dropping, and wind-blowing energy. The triboelectric performance of most textile TENGs easily fails under loading forces because of their weak mechanical properties (Dassanayaka et al., 2022).

4.3 Conclusions

Since the development in material science is endless, TENG reinforced with STF will be one of the progress in this field. There are very restricted numbers of STF-TENG producer institutes, especially with enhanced mechanical properties. Hence, this field is a new professional area for smart materials. On the other hand, commercialization of STF-TENG is not very common in the world. Putting a new product on market is quite easier since STF-TENGs reveal a ready-to-use product. Know-how is the most important point for the commercialization process. STF-TENG focuses on the use of novel TENG in protective applications and thereby aims at the defense industry at the first glance. However, TENG is known for energy harvesting and human motion monitoring as well. In this light, aerospace, medical, and electronics may be the target sectors related to the developed novel STF-TENG. One of the biggest aims of STF-TENG is to increase the mechanical properties of conventional TENGs by using STF and nanofillers. Therefore, by increasing the mechanics of this novel product, application areas of TENGs will be broadened from static systems to structures being subjected to dynamic loadings. This may lead to the replacement of conventional materials with TENG for reduced weight.

References

Dassanayaka, D. G., Alves, T. M., Wanasekara, N. D., Dharmasena, I. G., & Ventura, J. (2022). Recent progresses in wearable triboelectric nanogenerators. *Advanced Functional Materials, 32*, 2205438.

Wang, S., et al. (2020). Advanced triboelectric nanogenerator with multi-mode energy harvesting and anti-impact properties for smart glove and wearable e-textile. *Nano Energy, 78*, 105291.

Wang, W., et al. (2022). Enhanced Kevlar-based triboelectric nanogenerator with anti-impact and sensing performance towards wireless alarm system. *Nano Energy, 91*, 106657.

Yun, S.-Y., et al. (2022). Mechanically robust triboelectric nanogenerator with a shear thickening fluid for impact monitoring. *Journal of Materials Chemistry A, 10*, 10383–10390.

Chapter 5
A Short Review of Aircraft Noise Effects on Children's Learning in Auditory, Non-auditory, and Cognitive Development

Korawit Fakkhong, Sophita Sa-Ngobkai, Supansa Yamee, Tipavinee Suwanwong Rodbundith, and Nattaphon Rangsaritvorakarn

5.1 Introduction

Noise pollution causes annually at least 16,000 cases of premature death and over 72,000 hospitalisations in Europe (Lebrusán & Toutouh, 2020), and about 45,000 children (aged 7–19 years) are affected from noise exposure (Basner et al., 2014). Relevant studies show that the noise including within classroom noise, external noise, and environmental noise generated some negative impact directly on children including physical, mental health, and their learning ability (Beutel et al., 2016; Adewale et al., 2021). The external noise such as road traffic, aircraft, and industry were found to have a significant negative impact on both children's learning and academic performance at school (Stansfeld & Clark, 2015; Beutel et al., 2016; Wen et al., 2019).

The effects of noise on children at school particularly external noise from aircraft noise have been investigated in many studies for decade. The relevant studies show that the aircraft noise has a detrimental impact on the cognitive development of children (Beutel et al., 2016), physical and mental health (Stansfeld & Clark, 2015), and their life quality at school (Beutel et al., 2016; Adewale et al., 2021). These effects appear to be long term; the aircraft noise reduces the learning ability, where older children in high school are more affected when compared to younger children in primary school (Basner et al., 2014).

K. Fakkhong (✉) · T. S. Rodbundith · N. Rangsaritvorakarn
School of Management, Aviation Business Management, Healthcare Logistics Research Center, Mae Fah Luang University, Chiang rai, Thailand
e-mail: korawit.fak@mfu.ac.th; Tipavinee.suw@mfu.ac.th; Nattaphon.rag@mfu.ac.th

S. Sa-Ngobkai · S. Yamee
School of Management, Aviation Business Management, Mae Fah Luang University, Chiang rai, Thailand
e-mail: 6331210170@lamduan.mfu.ac.th; 6331210167@lamduan.mfu.ac.th

T. H. Karakoç et al. (eds.), *Energy and Sustainable Aviation Fuels Solutions*, Sustainable Aviation, https://doi.org/10.1007/978-3-031-70694-3_5

This chapter aimed to describe available evidence, research achievement, and remaining gaps of knowledge in the field of impacts of aircraft noise on children's learning. This chapter is chosen to focus on noise exposure and children's performance auditory and non-auditory, as well as effects of noise on children's cognitive development.

5.2 Systematic Reviews

This chapter gives a systematic review with three steps, (i) all studies were identified that could potentially be included in the review, (ii) then identified studies were submitted to selection criteria, (iii) the studies that meet the criteria were summarised. The areas covered include the impact of aircraft noise on children's performance in auditory tasks, non-auditory tasks, and children's cognitive development.

This chapter identified the studies published between January 2013 and January 2023 about the impact of noise exposure on children, (i) effect on children's performance in auditory tasks (ii) effect on children's performance in non-auditory tasks, and (iii) effect on children's cognitive development.

The publications that were identified were selected for data extraction. Three criteria were developed and implemented; (i) studies should report on the relationship between noise exposure particularly aircraft noise and children's learning or life quality at school and (ii) studies must quantify the sound levels, physical-mental health, learning ability, and relevant outcomes.

The studies that met the selection criteria were then summarised follows: (i) study data characteristics, (ii) population, (iii) exposure assessment, and (iv) results of the study.

5.3 Results and Discussion

Most studies reviewed so far were conducted with school-age children and focused on the impact of noise on children's learning. This section reports findings concerning effects of acute noise on children's performance auditory and non-auditory tasks, as well as effects of noise on children's cognitive development.

Effect of Noise on Children's Performance in Auditory Tasks
The question arises how noise saffects performance in listening tasks; listening in classroom is not only identification of a single word but requires semantic and syntactic processing of complex oral information while developing a story meaning. The relevant studies shown that unfavourable listening conditions such as external and environmental noise have a significant impact on children's speech perception and recognition accuracy (Klatte et al., 2013; Adewale et al., 2021). In addition, environmental noise can negatively affect both children's auditory working memory

and reading comprehension (Sullivan et al., 2015). Osman and Sullivan (2014) conducted an experiment with school-age children (8–10 years), where the children were asked to recall sequences of auditory stimuli presented in various levels of background noise. The results suggested that degraded listening conditions from background noise negatively impact the auditory working memory performance and listening comprehension. Similarly, Sullivan (2015) tested 63 children and 9 adults who were asked to recognise speech in various levels of background noise. The results suggested that background noise has a significant negative impact on speech recognition performance, and the study found that the negative impact of noise was more pronounced in children than in adults. To summarise, children's learning performance such as working memory, listening, and speech are associated with noise exposure (Klatte et al., 2013; Osman & Sullivan, 2014), and children require more favourable listening conditions than adults (Sullivan et al., 2015).

Effect of Noise on Children's Performance in Non-auditory Tasks
The relevant studies indicated that noise exposure is associated with several non-auditory health effects in children including annoyance, sleep disturbances, cognitive impairment, and cardiovascular and metabolic disorders (Basner et al., 2014; Waye & Kempen, 2021; Terzakis et al., 2022). Basner et al. (2014) mentioned that noise annoyance can affect daily activity, feelings, and sleep and children might end up with anger, displeasure, exhaustion, and stress-related symptoms. The studies have also shown that the noise exposure effect appears to be long term, noise effects have an interrelationship with the cardiovascular system, and causes relevant diseases—hypertension, heart diseases, and stroke. Terzakis et al. (2022) mentioned that sound pressure level and frequency were significantly associated with health effects in children. Fernandes et al. (2019) performed studies on children's performance in non-auditory tasks including reading and writing, where 162 children were randomly assigned to a quiet and noisy condition with sound pressure levels of 20 dB and 40 dB. The results showed that children in noisy conditions performed significantly worse on the reading and writing tasks, and its performance are associated with intense noise levels. To summarise, non-auditory tasks such as sleep and reading and writing abilities are also impaired by noise (Basner et al., 2014; Fernandes et al., 2019). Also, noise can have adverse health effects in the long term (Basner et al., 2014; Terzakis et al., 2022).

Effect of Noise on Children's Cognitive Development
The interesting question arises whether enduring exposure to environmental noise and external noise may cause persisting deficits in children's cognitive development. WHO suggests that the background noise should not exceed 35 dB during teaching sessions (Basner et al., 2014); reductions in noise are associated with improvements in cognition. The increasing levels of noise exposure due to road traffic, trains, aircraft, and construction, particularly in urban areas, are associated with cognitive performance in children (Basner et al., 2014; Stansfeld & Clark, 2015; Raess et al., 2022). Relevant studies show that aircraft noise generated a negative impact on children's attention (Stansfeld & Clark, 2015) and reduced memory (Stansfeld & Clark, 2015), but exposure to aircraft noise was consistently associated

with lower reading performance (Clark & Sörqvist, 2012). Similarly, Basner et al. (2014) conducted a study in which 2844 children aged 9–10 years at 89 schools around Heathrow airport (London, the United Kingdom) and Schiphol airport (Amsterdam, the Netherlands) were engaged in their research; the results suggested noise exposure, particularly aircraft noise associated with a 2-month delay in children's reading ability in the United Kingdom and a 1-month delay in the Netherlands. Raess et al. (2022) conducted a study in which the exposure to community noise levels of over 1200 children in Germany from birth to age 10 years was measured. The results suggested that higher levels of noise were associated with lower cognitive and behavioural development. They also mentioned that high levels of noise were linked to lower working memory, attention, and reading ability, which resulted in increased hyperactivity and emotional problems. In addition, Stansfeld and Clark (2015) mentioned that noise exposure affects physical and psychosocial behavior such as parental conflict, violence, and substance abuse, which result in lower cognitive development in children. To summarise, noise interference can influence learning ability, as well as worsen performance in reading and writing tasks at more intense noise levels (Basner et al., 2014; Fernandes et al., 2019; Raess et al., 2022).

5.4 Conclusion

The reviewed studies present empirical evidence substantiating the adverse effects of aircraft noise on children's learning. The results suggested that children are considerably more impaired by aircraft noise than adults, as well as older children are more affected when compared with younger children. Cognitive functions such as short memory, reading, and writing are also impaired by aircraft noise. The available evidence indicates that aircraft noise effects have to be taken seriously in view of possible long-term effects and the accumulation of risk factors in noise-exposed children. The finding reported providing further evidence of the detrimental impact of aircraft noise upon children at school, which required an appropriate acoustic design and collaboration between multi-sectors—government bodies, urban planners, and educators to minimise these effects.

References

Adewale, B., Jegede, F., Okubote, F., & Olagbadeguns M. (2021, March). Impact of classroom environments' on the academic performance of architecture students in Covenant University. In *IOP conference series: Earth and environmental science* (Vol. 665, No. 1, p. 012017). IOP Publishing.

Basner, M., Babisch, W., Davis, A., Brink, M., Clark, C., Janssen, S., & Stansfeld, S. (2014). Auditory and non-auditory effects of noise on health. *The Lancet, 383*(9925), 1325–1332.

Beutel, M. E., Jünger, C., Klein, E. M., Wild, P., Lackner, K., Blettner, M., & Münzel, T. (2016). Noise annoyance is associated with depression and anxiety in the general population-the contribution of aircraft noise. *PLoS One, 11*(5), e0155357.

Clark, C., & Sörqvist, P. (2012). A 3 year update on the influence of noise on performance and behavior. *Noise & Health, 14*, 292–296. https://doi.org/10.4103/1463-1741.104896

Fernandes, R. A., Vidor, D. C. G. M., & Oliveira, A. A. D. (2019). The effect of noise on attention and performance in reading and writing tasks. In *CoDAS* (Vol. 31). Sociedade Brasileira de Fonoaudiologia.

Klatte, M., Bergström, K., & Lachmann, T. (2013). Does noise affect learning? A short review on noise effects on cognitive performance in children. *Frontiers in Psychology, 578*.

Lebrusán, I., & Toutouh, J. (2020). Using smart city tools to evaluate the effectiveness of a low emissions zone in Spain: Madrid central. *Smart Cities, 3*(2), 456–478.

Osman, H., & Sullivan, J. R. (2014). Children's auditory working memory performance in degraded listening conditions. *Journal of Speech, Language, and Hearing Research, 57*(4), 1503–1511.

Raess, M., Brentani, A. V. M., Flückiger, B., de Campos, B. L. D. A., Fink, G., & Röösli, M. (2022). Association between community noise and children's cognitive and behavioral development: A prospective cohort study. *Environment International, 158*, 106961.

Stansfeld, S., & Clark, C. (2015). Health effects of noise exposure in children. *Current Environmental Health Reports, 2*, 171–178.

Sullivan, J. R., Osman, H., & Schafer, E. C. (2015). The effect of noise on the relationship between auditory working memory and comprehension in school-age children. *Journal of Speech, Language, and Hearing Research, 58*(3), 1043–1051.

Terzakis, M. E., Dohmen, M., van Kamp, I., & Hornikx, M. (2022). Noise indicators relating to non-auditory health effects in children—A systematic literature review. *International Journal of Environmental Research and Public Health, 19*(23), 15633.

Waye, K. P., & van Kempen, E. (2021). *Non-auditory effects of noise: An overview of the state of the science of the 2017–2020 period*. In Proceedings of the international commission on biological effects of noise.

Wen, X., Lu, G., Lv, K., Jin, M., Shi, X., Lu, F., & Zhao, D. (2019). Impacts of traffic noise on roadside secondary schools in a prototype large Chinese city. *Applied Acoustics, 151*, 153–163.

Chapter 6
Gaining of Conductivity in Shear Thickening Fluids

Selim Gürgen

6.1 Introduction

Shear thickening fluids have an increasing viscosity under increasing shear rates. Various models explain the shear thickening behavior touching upon the particle interactions in the microstructures. The hydro-clustering theory, the order-disorder theory and the contact rheology model are the most reliable models related to shear thickening formation. Apart from the physics of shear thickening fluids, researchers have been focused on the applications of these materials in engineering systems. For this purpose, shear thickening fluids have been integrated into many systems such as fabrics, foams, dampers, soft body armors and structures to enhance energy absorbing, vibration attenuation, shock suppression, etc. capabilities. Despite these applications, multi-functional system design has been a rising area to study, and therefore, shear thickening fluids have been investigated for multi-functional properties. One of the multi-functionalities is electrical conductivity in addition to the energy/vibration/shock absorbing properties. For this reason, researchers have made effort to make the shear thickening fluids multi-functional by gaining electrical conductivity in these suspensions. With the introduction of carbon-based conductive fillers such as carbon nanotubes (CNTs), shear thickening fluids get conductive because CNTs are good conductive materials for advanced engineering applications.

The rheological and conductive behaviors of CNT including the shear thickening fluids were studied by Wei et al. (2020). According to their study, CNT leads to significant shear thinning and thickening properties in the shear thickening fluids. Moreover, conductivity is enhanced by the effect of CNTs in the suspensions. Nakonieczna et al. (2019) fabricated shear thickening fluids including 0.25% CNTs. From this work, it is possible to mention that CNT inclusion enhances the

S. Gürgen (✉)
Department of Aeronautical Engineering, Eskişehir Osmangazi University, Eskişehir, Türkiye
e-mail: sgurgen@ogu.edu.tr

T. H. Karakoç et al. (eds.), *Energy and Sustainable Aviation Fuels Solutions*,
Sustainable Aviation, https://doi.org/10.1007/978-3-031-70694-3_6

rheological properties and is gaining electrical conductivity in the shear thickening fluids. Hasanzadeh et al. (2016) developed a shear thickening fluid system with a content of 0.8% CNT. These intelligent suspensions were suggested for wearable electronics including conductive shear thickening fluids with conductive properties. Chen et al. (2017) fabricated conductive shear thickening fluids with CNT addition. They found that as the CNT content is increased, the shear thickening effect is reinforced. Moreover, the electrical conductivity of the shear thickening fluids sharply increases by the effect of CNT addition.

6.2 Experimental Details

The shear thickening fluids were fabricated by dispersing a silica phase in a carrier liquid medium, polyethylene glycol. The mixture was blended for a half hour to ensure a homogenous microstructure. After the mixing process, the suspension was rested to remove the bubbles inside. The material was rheologically investigated in an MCR 302 rheometer by increasing the shear rate. To fabricate the conductive shear thickening fluids, CNT addition was used in the suspensions.

6.3 Results and Discussion

Figure 6.1 shows the rheological flow curves for two shear thickening fluids; one is bare shear thickening fluids and the other one includes CNT with a loading rate of 45%. From the chart, it is clearly seen that the suspensions show a drastic viscosity jump at around the shear rate of 25 s^{-1}. The bare shear thickening fluid has a peak

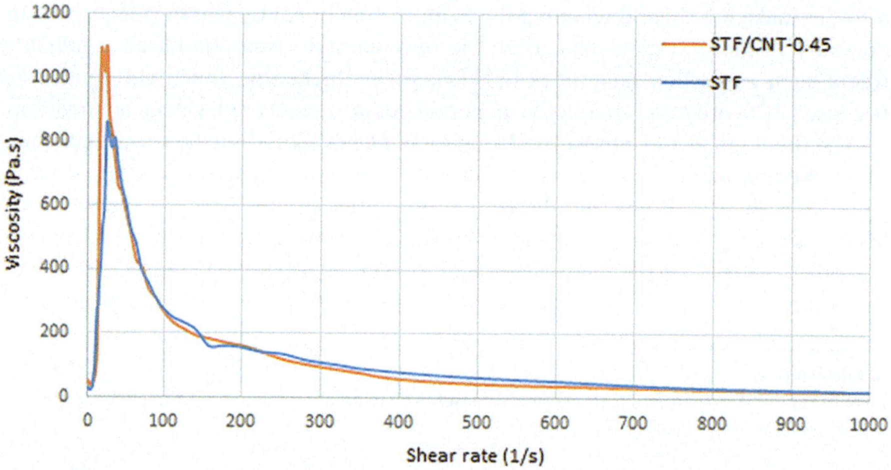

Fig. 6.1 Flow curves for the samples

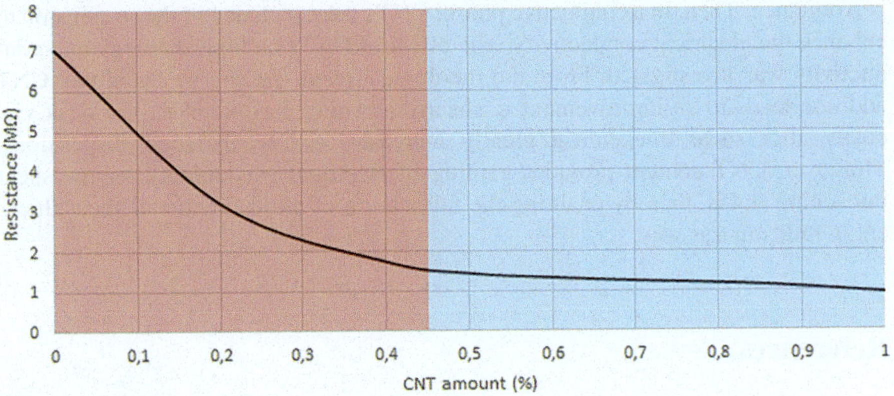

Fig. 6.2 Electrical resistance for the samples

viscosity of 855 Pa.s, whereas the one including CNT reaches a peak viscosity of 1100 Pa.s. Hence, it is possible to mention that CNT content enhances the shear thickening properties, thereby increasing the maximum viscosity levels. It is also important to state that CNT inclusion leads to a much sharper viscosity jump at the critical shear rate because the viscosity climb has a higher jump angle with the CNT-included suspension. After reaching the maximum viscosity point, the suspensions show a slight reduction in the viscosities, which are called shear thinning. This is because of the particle-network breakdown in the microstructure of the suspensions. Beyond the peak viscosity point, the particles within the suspensions cannot bear the shear stresses developed in the mixtures.

Figure 6.2 shows the electrical resistances for the shear thickening fluids including different amounts of CNT. From the results, the bare shear thickening fluid has the maximum resistance value at about 7 MΩ because the suspension without CNT is not capable of electrical conductivity. By including CNT in the suspensions, conductivity is gradually developed, thereby reducing the electrical resistance. There is a threshold at about 45% of CNT content in the shear thickening fluid. After this point, the electrical reissuance in the shear thickening fluids is quite stable. For this reason, percolation is at the CNT amount of 45% for the shear thickening fluid investigated in this study. According to the results, it can be stated that CNT inclusion provides gaining of electrical conductivity for the shear thickening fluids, thereby enabling the fabrication of the multi-functional shear thickening fluids by this way.

6.4 Conclusions

In the present work, a conventional shear thickening fluid was fabricated by using nano-size silica particles distributed in a polyethylene glycol medium. The rheology of this material was investigated under increasing shear rates. Moreover, electrical conductivity measurements were carried out to understand the conductive performance. Then, in a conductive phase, CNTs were included in the suspension to enhance the electrical conductivity. The effect of CNTs on both rheology and conductivity was investigated. From the rheological results, it can be stated that CNT addition leads to an improvement in shear thickening performance. The peak viscosity after shear thickeninsg greatly increases in CNT-included suspensions. Moreover, CNT content provides gaining of electrical conductivity for the shear thickening fluids, thereby enabling the fabrication of multi-functional shear thickening fluids in this way.

References

Chen, Q., Liu, M., Xuan, S., Jiang, W., Cao, S., & Gong, X. (2017). Shear dependent electrical property of conductive shear thickening fluid. *Materials & Design, 121*, 92–100.

Hasanzadeh, M., Mottaghitalab, V., Babaei, H., & Rezaei, M. (2016). The influence of carbon nanotubes on quasi-static puncture resistance and yarn pull-out behavior of shear-thickening fluids (STFs) impregnated woven fabrics. *Composites Part A: Applied Science and Manufacturing, 88*, 263–271.

Nakonieczna, P., Wierzbicki, Ł., Wróblewski, R., Płociński, T., & Leonowicz, M. (2019). The influence of carbon nanotube addition on the properties of shear thickening fluid. *Bulletin of Materials Science, 42*(4), 1–4.

Wei, M., Lv, Y., Sun, L., & Sun, H. (2020). Rheological properties of multi-walled carbon nanotubes/silica shear thickening fluid suspensions. *Colloid and Polymer Science, 298*(3), 243–250.

Chapter 7
High-Impact Resistance with Aerogel-Based Composites

Selim Gürgen

7.1 Introduction

Throughout history, armor systems have been developed to protect individuals from weaponry in combat. Body armor aims to provide the wearer with resistance to specific threats while maintaining mobility and minimizing bulk. All body armor systems require a balance between ballistic protection and user mobility (David et al., 2009). Aerogels are among the lowest-density solids in the world and are often referred to as frozen smoke (Tillotson & Hrubesh, 1992). They are made up of more than 90% space and have a finely structured porous skeletal framework that gives them various properties such as high surface areas, low thermal conductivities and dielectric constants, and high acoustic attenuation. Silica aerogels were the first aerogels synthesized by Kistler in the 1930s using a process that replaced the solvent in wet gels with air to create a supercritical fluid. However, silica aerogels have poor mechanical properties and are fragile due to their narrow interparticle necks. To overcome this, researchers have crosslinked silica aerogels with organic polymers to reinforce the interparticle necks and improve the overall mechanical strength (Ann et al., 2007). These crosslinked aerogels are known as polymer-crosslinked aerogels or X-aerogels and have been found to be exceptionally strong compared to non-crosslinked aerogels and other materials (Sudhir et al., 2008). Isocyanates, commonly used in the synthesis of polyurethanes and polyureas, are the most widely used crosslinkers for X-aerogels. Researchers have also explored the possibility of creating purely polymeric aerogels with similar mechanical properties as X-aerogels but with a simpler synthetic protocol (McGrath & James, 1977).

S. Gürgen (✉)
Department of Aeronautical Engineering, Eskişehir Osmangazi University, Eskişehir, Türkiye
e-mail: sgurgen@ogu.edu.tr

T. H. Karakoç et al. (eds.), *Energy and Sustainable Aviation Fuels Solutions*, Sustainable Aviation, https://doi.org/10.1007/978-3-031-70694-3_7

7.2 Literature Survey

The defense industry has always been focused on protecting military and civilian forces, and armor materials have played a crucial role in this regard. Researchers in defense technology strive to create lightweight body armor that can provide protection against close-range attacks, which are among the most dangerous and common types of threats. While durability is important, weight is also a critical factor in the design of body armor. Over time, numerous materials have been tested and used as armor, with ongoing development in response to advancements in technology and evolving threats. The armor systems used must offer sufficient protection against bullets, shrapnel fragments, and mine explosions while also meeting criteria such as low weight, high mobility, and continuity. The design of armor has also focused on the need for materials that are durable, lightweight, long-lasting, and highly mobile. As a result, single-layer and multilayer armor systems have been developed, with single-layer armor preferred for resistance to small-caliber bullets and multilayer armor for protection against armor-piercing bullets (Özgültekin, 2012).

The study by Capadona et al. (2006) investigated the effects of crosslinking on the mechanical properties of aerogels by systematically varying the silane concentration, di-isocyanate concentration, and polymerization temperature during preparation. The researchers prepared 28 different aerogel monoliths and found that the highest density crosslinked aerogel had the highest stress at failure and modulus. The stress at failure was 120 times higher than that of non-crosslinked aerogels, despite the density increasing by a factor of three. In contrast, the lowest density crosslinked aerogels showed a 15-fold increase in stress at failure with only a ~1.3 increase in density.

Polyurea-crosslinked, ordered mesoporous silica aerogels have been shown to have the potential for energy absorption applications. Churu et al. (2015) produced these aerogels using a sol-gel process with a three-block pluronic copolymer P123 and trimethylbenzene (TMB) as the structure-guiding and swelling agents, respectively. The size, shape, and distribution of the micelles in solution varied by changing the TMB concentration, resulting in different morphologies and mechanical and physical properties of the aerogels. The effect of TMB and crosslinking agent concentration on the mechanical behavior was investigated, and a honeycomb-like structure with the designated formula X-MP4-T045 was obtained, which showed a specific energy absorption of 123 J/g. This high energy absorption capacity makes it a promising material for applications in ballistic protection such as helmets, body armor, vehicles, and personnel protective equipment.

Katti et al. (2006) conducted a study on isocyanate crosslinked amine-modified silica aerogels to investigate their chemical, physical, and mechanical properties. They found that by encapsulating the skeletal framework of amine-modified silica aerogels with polyurea, they could create a lightweight composite material that was mechanically strong. The potential of the new crosslinked silica aerogels for load-carrying applications was tested through compression, three-point bending, and dynamic mechanical analysis. Silica aerogel is a low-density, chemically inert material with excellent acoustic insulation properties and low thermal conductivity, making it suitable for use in various applications such as energy storage, sensing, air/water purification, catalysis, and aerospace applications. Despite its beneficial properties, the fragility of aerogels limits their application to certain areas, such as Cerenkov radiation detectors in nuclear reactors, NASA's stardust program, and ultra-lightweight thermal insulation for planetary vehicles such as the Sojourner rover and the Spirit and Opportunity in 2004.

Luo et al. (2006) conducted research on the compressive behavior of crosslinked silica aerogel (CSA) using a split Hopkinson pressure bar. CSA is a type of aerogel that is denser and stronger than native aerogels due to polymer crosslinking. The researchers determined the stress-strain relationship of dry CSA at high strain rates and observed its deformation and failure behavior using high-speed photography and digital image correlation. The study found that CSA is more than 300 times stronger and less hydrophilic than native aerogels.

7.3 Conclusions

Advanced engineering materials with unique properties such as high mechanical strength, impact resistance, low density, and low cost are gaining attention in aerospace and military research. The ballistic performance of body armor has been continuously evaluated. Cross-linked polymer aerogels, which have high specific region, high porosity, low density, and excellent mechanical behavior, are considered new-generation materials with wide-ranging applications in aerospace research. These aerogels are the preferred choice for impact-resistant applications due to their ability to absorb kinetic energy. Although several groups and researchers have attempted to reinforce polymer aerogels, few studies have focused on producing aerogel-based textile composites for ballistic purposes.

References

Ann, B., et al. (2007). Structure-property relationships in porous 3D nanostructures as a function of preparation conditions: Isocyanate cross-linked silica aerogels. *Chemistry of Materials, 19*, 2247–2260.

Capadona, L. A., Meador, A. B., Alunni, A., Fabrizio, E. F., Vassilaras, P., & Leventis, N. (2006). Flexible, low-density polymer crosslinked silica aerogels. *Polymer, 47*, 5754–5761.

Churu, G., Zupančič, B., Mohite, D., Wisner, C., Luo, H., Emri, I., Leventis, C., Leventis, N., & Lu, H. (2015). Synthesis and mechanical characterization of mechanically strong, polyurea-crosslinked, ordered mesoporous silica aerogels. *Journal of Sol-Gel Science and Technology, 75*, 98–123.

David, N. V., Gao, X. L., & Zheng, J. Q. (2009). Ballistic resistant body armor: Contemporary and prospective materials and related protection mechanisms. *Applied Mechanics Reviews, 050802*, 1–20.

Katti, A., Shimpi, N., Roy, S., Lu, H., Fabrizio, E. F., Dass, A., Capadona, L. A., & Leventis, N. (2006). Chemical, physical, and mechanical characterization of isocyanate cross-linked amine-modified silica aerogels. *Chemistry of Materials, 18*, 285–296.

Luo, H., Lu, H., & Leventis, N. (2006). The compressive behavior of isocyanate-crosslinked silica aerogel at high strain rates. *Mechanics of Time-Dependent Materials, 10*, 83–111.

McGrath, J. E., & James, E. (1977). *Block copolymers: Overview and critical survey.* Academic Press.

Özgültekin, S. E. (2012). *Balistik zırhlarda kullanılan kompozit malzeme kombinasyonlarının incelenmesi*. Sakarya Üniversitesi.

Sudhir, et al. (2008). Cross-linking 3D assemblies of nanoparticles into mechanically strong aerogels by surface-initiated free-radical polymerization. *Chemistry of Materials, 20*, 5035–5046.

Tillotson, T. M., & Hrubesh, L. W. (1992). Transparent ultralow-density silica aerogels prepared by a two-step sol-gel process. *Journal of Non-Crystalline Solids, 145*, 44–50.

Chapter 8
Theoretical Performance Analysis of High-Entropy Alloys in Hybrid Rocket Motors

Muhammad Ezamuddin Ezzad Sabri, Muhammad Hanafi Azami, and Norhuda Hidayah Nordin

Nomenclature

CEA	Chemical equilibrium with applications
HEA	High-entropy alloys
HRM	Hybrid rocket motor
NASA	National Aeronautics and Space Administration
O/F	Oxidiser-to-fuel ratio
H_2O_2	Hydrogen peroxide
IRFNA	Inhibited red fuming nitric acid
N_2O	Nitrous oxide
N_2O_4	Nitrogen tetroxide
O_2	Gaseous oxygen
$O_2(L)$	Liquid oxygen

8.1 Introduction

A hybrid rocket motor's fuel mix can be either liquid or gaseous oxidant and solid fuel or the other way around (Sutton & Biblarz, 2016). Meeting the demands of missions with variable thrust, serial firing, and storage of non-toxic propellants is the key benefit over solid systems. In contrast to solid and liquid propellant rockets, hybrid rockets suffer from several propulsion efficiency drawbacks, including

M. E. E. Sabri · M. H. Azami (✉)
Department of Mechanical & Aerospace Engineering, Kulliyyah of Engineering, International Islamic University Malaysia, Gombak, Kuala Lumpur, Malaysia
e-mail: hanafiazami@iium.edu.my

N. H. Nordin
Department of Manufacturing & Materials Engineering, Kulliyyah of Engineering, International Islamic University Malaysia, Gombak, Kuala Lumpur, Malaysia

T. H. Karakoç et al. (eds.), *Energy and Sustainable Aviation Fuels Solutions*, Sustainable Aviation, https://doi.org/10.1007/978-3-031-70694-3_8

reduced propulsion efficiency and time-varying oxidiser-to-fuel ratio (O/F) during firing and throttling operations (Chiaverini & Kuo, 2007). The chemical characteristics of the energetic additions can be utilised to boost HRM's propelling effectiveness. By adding high-energy fuel components like metal and metal hydrides into the solid fuel grain, this method raises the regression rate.

High-entropy alloys (HEAs) are alloys made up of five or more principal components. Each key component should have a concentration between 5% and 35%. (Tsai & Yeh, 2014). HEA may also contain minor elements, each with a percentage below 5%, in addition to the core constituents. Because they exhibit substantially higher mixing entropies in their liquid or random solid solution phases than ordinary alloys, these alloys are referred to as "HEA." Therefore, the effects of entropy are considerably more apparent in HEA. Based on four major effects, this feature enables appropriate alloying elements to improve the materials' qualities. The four primary effects of HEA—high-entropy effects, slow diffusion, lattice distortion, and cocktail effects—come from the presence of numerous major elements. They provide significant potential for the utilisation of energetic particles that will influence the regression behaviour of hybrid rocket fuel (Yeh et al., 2016). The performance of HEA as additive materials for hybrid rocket propulsion systems is the main emphasis of this work overall.

8.2 Methodology

Various HEA concentrations and oxidisers were evaluated in this study. Table 8.1 lists the many HEAs that are employed. These HEAs were tested for how concentration in paraffin wax affected the metrics of propulsive performance. Oxidants such as O_2, O_2(L), N_2O, H_2O_2, N_2O_4, and inhibited red fuming nitric acid (IRFNA) were mixed with these fuels. The test was also conducted with varied O/F and chamber pressure to investigate their impact on the parameters governing propulsive performance. The free CEA-NASA software was used to do thermochemical simulations. A variety of assumptions are made during the computation of the performance characteristics for rocket motors, including:

- Zero velocity at the inlet of the combustion chamber.
- Homogeneous mixture.
- Complete combustion.
- No heat loss.
- One-dimensional flow.
- Isentropic expansion of the gases along the nozzle.

Table 8.1 Composition of $FeCoNiAl_{(1-x)}B_xSi$ HEA powders

	Composition (wt.%)					
Sample	Fe	Co	Ni	Al	B	Si
HEA1	0.27	0.27	0.27	0.05	0.05	0.10
HEA2	0.27	0.27	0.27	0.00	0.10	0.10

Table 8.2 Test conditions

Test	O/F	HEA concentration [wt.%]	HEA	Oxidiser
1	2	1–10	HEA1,2	O_2
2	1.0–3.0	1–10	HEA2	O_2
3	1.0–3.0	1	HEA2	O_2, O_2(L), N_2O, H_2O_2, N_2O_4, IRFNA

Table 8.3 Comparison of results of CEA-NASA and experiment

HEA composition [wt.%]	CEA thrust [N]	Experimental thrust [N]	Percentage error [%]
0	58.09	45.15	22.27
3	71.00	76.09	7.18
5	58.64	59.36	1.22
7	58.44	51.62	11.67

Characteristic velocity, specific impulse, and adiabatic flame temperature are the propulsive parameters considered in this project's context. The equilibrium condition was selected for this work because it represented an appropriate approximation to determine the upper limits of performances anticipated for the examined propellants. There were three separate test sets: (1) HEA of different compositions, (2) HEA concentrations between 1 and 10 wt.% with O/F values between 1 and 3, and (3) various oxidisers. All the simulations were done at 344.74 kPa in the combustion chamber with paraffin wax as the base fuel. The criteria for each test are shown in Table 8.2.

8.3 Results

The results from the simulation were compared to the experimental results, as shown in Table 8.3.

Both results indicate a similar pattern in which 3% HEA provides the most thrust. Table 8.4 shows the performance difference between two different HEA compositions. The outcome may be seen in the parameters as follows: specific impulse (Isp), characteristic velocity (c*), and adiabatic flame temperature (K).

In every scenario, HEA1 with aluminium had somewhat worse performance than HEA2. Regarding characteristic velocity, specific impulse, and adiabatic flame temperature, HEA2 has an average improvement of 0.04%, 0.05%, and 0.03%, respectively. This might be because HEA1, with a total latent heat of 126.9 J/g, is less thermally active than HEA2, which has a total latent heat of 136.34 J/g (Mohamad Jamil et al., 2023). Figure 8.1 shows the ideal values of characteristic velocity, specific impulse, and adiabatic flame temperature that the thermochemical simulation produced for all HEA concentrations.

Every HEA concentration has a maximum characteristic velocity when the O/F is 2.0. The fuel with 0% HEA had the highest average characteristic velocity (1752.6 m/s), whereas fuel with 10% HEA had the lowest. After reaching an O/F of

Table 8.4 Comparison of HEA1 and HEA2 with varying wt.% in paraffin wax

	c* [m/s]		I_{sp} [s]		Temperature [K]	
Concentration (wt.%)	*HEA1*	*HEA2*	*HEA1*	*HEA2*	*HEA1*	*HEA2*
1	1749.1	1749.2	117.08	117.09	3162.82	3162.95
2	1745.5	1745.8	116.76	116.78	3168.76	3169.03
3	1741.8	1742.2	116.42	116.45	3174.41	3174.80
4	1738.0	1738.5	116.09	116.12	3179.8	3180.39
5	1734.1	1734.7	115.75	115.80	3184.91	3185.68
6	1730.0	1730.8	115.40	115.46	3189.74	3190.71
7	1725.9	1726.8	115.06	115.13	3194.29	3195.48
8	1721.6	1722.7	114.70	114.78	3198.55	3199.97
9	1717.2	1718.5	114.34	114.43	3202.54	3204.21
10	1712.8	1714.2	113.99	114.09	3206.24	3208.17

*It shows the speed at which combustion products would leave the combustion chamber in the case that they passed through the ideal, hypothetical throat area without experiencing any losses

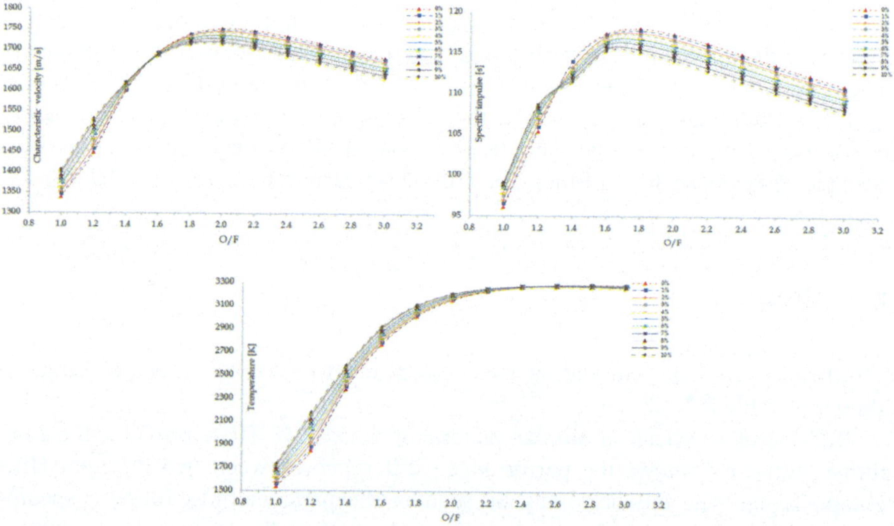

Fig. 8.1 Characteristic velocity, specific impulse, and adiabatic flame temperature of the concentrations of HEA2 as a function of O/F

1.4–1.6, the characteristic velocity of a higher HEA concentration begins to fall faster than that of a lower HEA concentration. For instance, during 1.0–1.4, the characteristic velocity of 10% HEA was higher than that of 9% HEA, but at 1.6–3.0, it dropped below the 9% HEA.

When the O/F is at 1.8, the specific impulse is maximum for all HEA concentrations. The fuel containing 0% HEA had the highest peak at 118.05 s, whereas the fuel containing 10% HEA had the lowest peak at 115.09 s. The specific impulse follows the same trajectory as the characteristic velocity, starting higher than the

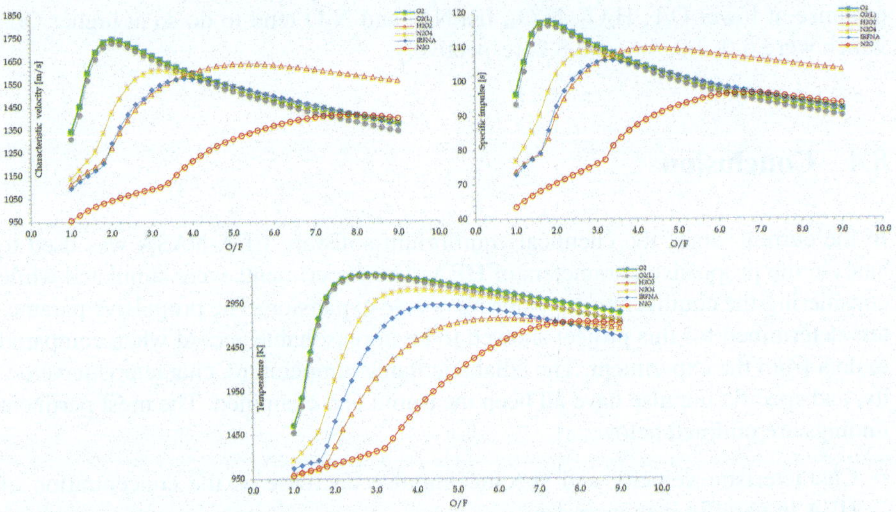

Fig. 8.2 Characteristic velocity, specific impulse, and adiabatic flame temperature of the different oxidants as a function of O/F

specific impulse of a lower HEA concentration but declining quickly after reaching an O/F of 1.4.

When the O/F is between 2.6 and 2.8, the highest adiabatic flame temperature is reached for each HEA concentration. Although it has the lowest average temperature of 2825.02 K, the fuel with 0% HEA has the greatest peak temperature of 3291.4 K. The fuel containing 10% HEA had the highest average temperature of 2916.36 K but the lowest peak temperature of 3277.1 K. Although it happens later in the O/F, the adiabatic flame temperature trended in the same direction as the other metrics. Figure 8.2 shows the optimal values of characteristic velocity, specific impulse, and adiabatic flame temperature discovered using thermochemical modelling for all oxidants.

When O_2 is used as the oxidiser, the characteristic velocity is at its greatest, 1729 m/s, for an O/F of 2.0. The lowest peak value, however, was obtained when N_2O was used as an oxidiser which was 1415.6 at an O/F of 7.8. The composition of combustion products that contain nitrogen which does not participate in the oxidation process while retaining some of the generated heat should be the primary cause.

With an O/F of 1.8, O_2 exhibits the highest value of specific impulse, with a peak of 117.79 s. The lowest average specific impulse value comes from N_2O, but it becomes higher than most oxidisers when reaching an O/F of 7.0. N_2O_4 and IRFNA follow the same trend as O_2 and O_2(L) after reaching the peak value.

When O_2 is used as the oxidiser, the highest temperature occurs at 3290.2 K with an O/F ratio of 2.8. Similarly, O_2(L) yields temperature data with the same trend curve but a lower peak of 3261.2 K for the identical O/F. Up to an O/F of 1.8, the adiabatic temperatures of H_2O_2 and N_2O are comparable. While O_2 and O_2(L) optimise at lower O/F, H_2O_2, N_2O_4, IRFNA, and N_2O tend to do so at higher O/F, which were 6.0, 4.0, 4.6, and 8.2, respectively.

8.4 Conclusion

In the current work, the chemical equilibrium software CEA-NASA was used to analyse the propulsive parameters of HEA. The experiments were simulated while considering the equilibrium state during nozzle expansion. The propulsive parameters determined for this project showed to be in reasonable accord when compared to data from the experiment. The adiabatic flame temperature, characteristic velocity, and specific impulse have all been measured and examined. The most pertinent findings are outlined below:

- Characteristic velocity and specific impulse decrease as the concentration of HEA in paraffin wax increases.
- A higher HEA percentage increases the average adiabatic flame temperature but has a lower peak temperature.
- O_2 presents the highest values for all the propulsive parameters.
- N_2O presents the lowest values of the performance parameters.

Setting more precise initial conditions will allow further advancements in this study's data accuracy. To improve the representation of the experimental data, the HEA and oxidiser properties can also be discussed in greater detail.

Acknowledgement The authors want to thank and acknowledge the Asian Office of Aerospace R&D (AOARD) and the Ministry of Higher Education Malaysia (MOHE). This research is supported by AOARD Grant (SPI21-112-0112) and FRGS Grant (FRGS/1/2021/TK0/UIAM/03/2).

References

Chiaverini, M. J., & Kuo, K. K. (2007). *Fundamentals of hybrid rocket combustion and propulsion*. International American Institute of Aeronautics and Astronautics.

Mohamad Jamil, N. A., Mohamaddiah, A. H., Azami, M. H., & Nordin, N. H. (2023). *Materials today : Proceedings high entropy alloy as metal additive for hybrid rocket propellant*. In Materials today: Proceedings, xxxx. https://doi.org/10.1016/j.matpr.2022.10.257.

Sutton, G. P., & Biblarz, O. (2016). *Rocket propulsion elements*. Wiley.

Tsai, M. H., & Yeh, J. W. (2014). High-entropy alloys: A critical review. *Materials Research Letters, 2*(3), 107–123. https://doi.org/10.1080/21663831.2014.912690

Yeh, M. C., Jien-wei, G., Liaw, P. K., & Zhang, Y. (2016). *High-entropy alloys, fundamentals and application*.

Chapter 9
Selection of Sustainable Aviation Fuels: An Expert-Based Comparative Approach

Omar Alharasees and Utku Kale

Nomenclature

AHP	Analytical hierarchy process
MCDM	Multi-criteria decision-making
PCM	Pairwise comparison matrix
CR	Consistency ratio
SAF	Sustainable aviation fuel

9.1 Introduction

Aviation, a prominent contributor to greenhouse gas emissions, poses a formidable challenge in mitigating its environmental impact (Perea-Moreno et al., 2022). Sustainable aviation fuel emerges as a promising solution to address this concern, enabling the aviation industry to reduce its carbon footprint while accommodating the escalating fuel demand (Bergero et al., 2023), However, the development of sustainable aviation fuel (SAF) is influenced by various factors, encompassing economic, technical, and social acceptability considerations.

SAF faces cost-related challenges, exceeding conventional fuels due to limited feedstock and costly conversion technologies (del Monte et al., 2022). The absence of regulatory frameworks and carbon pricing mechanisms further hampers SAF's progress (Detsios et al., 2023). Ensuring technical feasibility is crucial (Shahriar & Khanal, 2022). Compatibility with existing aviation infrastructure, including aircraft, engines, and fuelling systems, is essential. Meeting fuel specifications for

O. Alharasees (✉) · U. Kale
Department of Aeronautics and Naval Architecture, Faculty of Transportation Engineering and Vehicle Engineering, Budapest University of Technology and Economics, Budapest, Hungary
e-mail: oalharasees@edu.bme.hu; kale.utku@kjk.bme.hu

T. H. Karakoç et al. (eds.), *Energy and Sustainable Aviation Fuels Solutions*, Sustainable Aviation, https://doi.org/10.1007/978-3-031-70694-3_9

energy density and combustion characteristics ensures safety and performance. Modifications to storage and handling may incur additional adoption costs.

Social acceptability significantly influences SAF advancement (Anderson et al., 2022). Public perception and acceptance drive industry investment. Health impacts, community engagement, and involvement must be considered (Xu et al., 2022). SAF's development is pivotal for mitigating aviation's environmental impact. Economic, technical, and social factors influence SAF's adoption, necessitating collaboration among industry, policymakers, and the public (Ahmad et al., 2019).

Despite numerous advantages, SAFs face significant obstacles. Foremost is the production cost, surpassing conventional jet fuel (Tanzil et al., 2021) due to feedstock prices, conversion technologies, and economies of scale. However, advancing technology and increased production can lead to price reduction. Limited feedstock availability poses a challenge in sustainable aviation fuel (SAF) production (O'malley et al., 2021). Factors like agricultural productivity, weather patterns, and land use policies influence feedstock accessibility and cost (Mofijur et al., 2023). Concerns also arise regarding certain feedstocks, particularly food crops, due to potential conflicts with food security and competing land uses, such as conservation efforts. This study employs an analytical hierarchy process (AHP) to assess the preferences of experienced (over 3 years of experience) and less-experienced experts (under 3 years of experience) towards SAF. The evaluation considers participant experience levels, categorizing them into two groups. A hierarchical model is constructed using key criteria, facilitating multi-level decision-making and addressing conflicts and uncertainties. The study aims to provide insights for informed decision-making and policy shaping in the aviation industry, supporting a sustainable transition to low-carbon aviation.

9.2 Method

Thomas L. Saaty developed the analytical hierarchy process (AHP) in the 1980s to facilitate systematic evaluation and decision-making. AHP integrates both mathematical and psychological principles (Saaty, 1990) as outlined by Saaty in his works published in 1990 and 1994. The methodology employs objective mathematical techniques to assist individuals or groups in re-evaluating their priorities following a decision-making process. By using AHP, individuals or groups can effectively analyse and evaluate complex decision problems, allowing them to arrive at well-informed and sound decisions (Saaty, 1994). The systematic AHP approach to decision-making challenges is concisely outlined in Fig. 9.1.

AHP is a technique utilized for the evaluation and prioritization of options through the establishment of standards. By assigning percentages to the options based on their relative importance or influence, AHP enables the creation of a decision hierarchy. This, in turn, facilitates the prediction and decision-making process by establishing a clear order of preference for the options being evaluated. As such, AHP is an effective tool for complex decision-making problems where a hierarchy

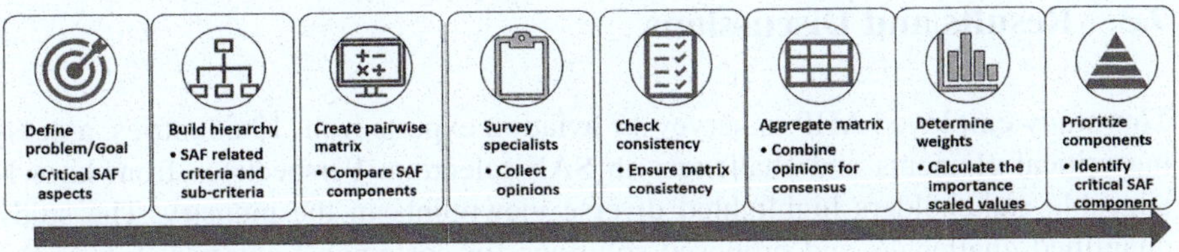

Fig. 9.1 AHP method steps

Table 9.1 Saaty scale (Saaty, 1977)

Mathematical representation	1	2,4,6,8	3	5	7	9
Meaning	Both are equally important	Values between moderate and strong importance	One is moderately more important than the other	One is strongly more important than the other	One is very strongly more important	One is extremely more important than the other

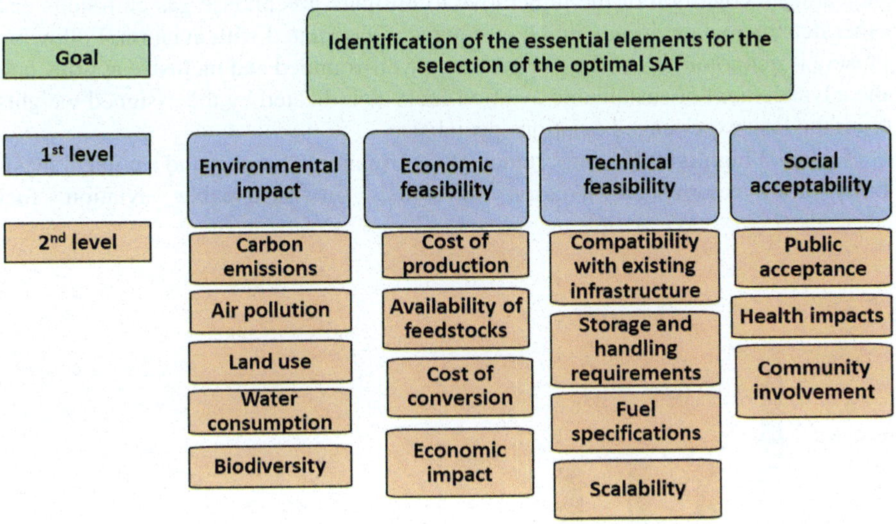

Fig. 9.2 The hierarchal model

of options must be established and evaluated. AHP employs individual benchmarks and the Saaty scale to rank decisions as shown in Table 9.1.

Incorporating diverse decision-influencing factors and quantifying the relative importance of selected decisions compared to other model elements helps ensure a comprehensive analysis (Saaty, 1977, 2002). This enables accurate capture and expression of variations in importance as quantitative percentages. The AHP method effectively identifies critical components impacting SAF implementation in aviation (Fig. 9.2).

9.3 Results and Discussions

This study employed AHP to survey 17 aviation experts from 12 countries, assessing critical elements and challenges in SAF selection. Perspectives from knowledgeable stakeholders highlighted diverse viewpoints in the industry. The study quantified challenges and proposed solutions for sustainable aviation. Figure 9.3 illustrates the distribution of participating countries.

Analysing respondents' opinions on the hierarchical approach to SAF evaluation reveals divergent viewpoints based on varying expertise levels, occupational classifications, and comprehension of the current situation and its progression. However, employing AHP for pairwise comparison provides a comprehensive perspective on the development of sustainable aviation fuel within the contemporary system, surpassing simplistic techniques. Using the designated methodology, initial grouping was conducted according to experience classifications and averaged responses were computed using the geometric mean. Each group calculated weights and consistency ratios for the first level in the SAF model, which are presented in Tables 9.2 and 9.3.

Table 9.2 displays the criteria weights for evaluating SAF as perceived by less-experienced aviation specialists. Environmental impact receives the highest weight, indicating recognition of the imperative to mitigate greenhouse gas emissions and other detrimental environmental consequences associated with aviation. Following closely is economic feasibility, indicating a well-rounded and inclusive approach to the advancement of sustainable aviation fuels as indicated by the assigned weights from the less-experienced aviation specialists.

Table 9.3 highlights the importance of environmental impact and technical feasibility for experienced aviation specialists in sustainable aviation fuel

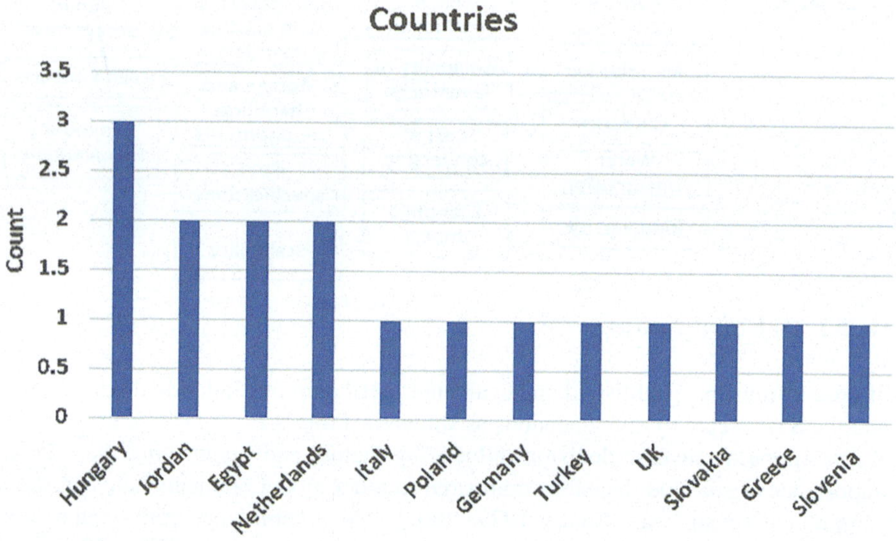

Fig. 9.3 The distribution of the participating countries

Table 9.2 Less-experienced aviation specialists PCMs for the first level

LESS EXPERIENCED					
Criteria	Environmental impact	Economic feasibility	Technical feasibility	Social acceptability	Weights
Environmental impact	1.00	1.47	1.28	1.52	31.63%
Economic feasibility	0.68	1.00	2.06	1.64	30.13%
Technical feasibility	0.78	0.49	1.00	1.45	20.89%
Social acceptability	0.66	0.61	0.69	1.00	17.36%
CR=0.0311		Sum=			100%

Table 9.3 Experienced aviation specialist PCMs for the first level

WELL EXPERIENCED					
Criteria	Environmental impact	Economic feasibility	Technical feasibility	Social acceptability	Weights
Environmental impact	1.00	0.77	1.49	1.51	28.68%
Economic feasibility	1.29	1.00	0.72	1.06	24.90%
Technical feasibility	0.67	1.39	1.00	1.51	26.93%
Social acceptability	0.66	0.94	0.66	1.00	19.49%
CR=0.0347		Sum=			100%

decision-making. Economic feasibility and social acceptability have lower significance. Prioritizing sustainability in aviation is crucial, and sustainable aviation fuels (SAFs) are seen as a key step towards carbon neutrality by 2050 (Abrantes et al., 2021), and the adoption of SAF is seen as a critical step in achieving this goal. For a comprehensive analysis of survey results and critical elements in the current model, a thorough comparison of all aspects from different expert perspectives is crucial. Figure 9.4 presents a comprehensive comparison of aviation groups, considering participant experience levels.

Environmental impact was identified as the foremost criterion in both groups, with economic feasibility following suit. Notably, experienced aviation experts exhibited a higher emphasis on technical feasibility relative to economic feasibility. These variations in priority underscore the significance of employing the AHP method, which allows for the consideration of diverse perspectives.

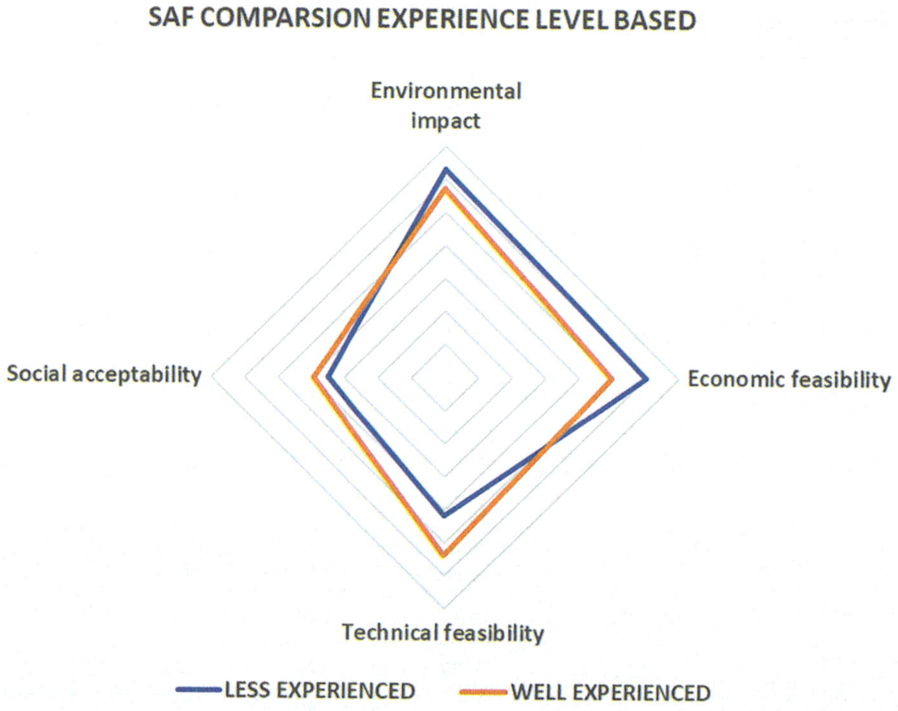

Fig. 9.4 SAF-AHP criteria comparison—experience based

9.4 Conclusion

Sustainable aviation fuels (SAFs) promise to reduce the aviation industry's environmental impact. Selecting the optimal SAF alternative requires evaluating economic feasibility, technical feasibility, and social acceptability. This study employed a multi-criteria decision-making (MCDM) approach using the analytical hierarchy process (AHP) method. AHP prioritized components, aiding aviation decision-makers from different levels of expertise in harmonizing stakeholders' objectives.

Findings highlight the importance of evaluating environmental impact, specifically carbon emissions. Economic feasibility and technical feasibility are influential factors. Policymakers can utilize these findings to identify aligned SAF alternatives. Challenges include production costs, limited feedstock availability, expensive conversion technologies, regulatory ambiguity, and the absence of a carbon pricing mechanism. As technology advances and production scales up, SAF prices are expected to decrease. Technical feasibility ensures safety and performance. Infrastructure modifications may increase adoption costs.

In conclusion, collaboration among industry, policymakers, and public drives SAF development. AHP empowers decision-makers to prioritize critical factors. SAF adoption is crucial for global climate targets. This study's insights benefit policymakers and stakeholders committed to sustainable aviation.

References

Abrantes, I., et al. (2021). Sustainable aviation fuels and imminent technologies—CO2 emissions evolution towards 2050. *Journal of Cleaner Production, 313*, 127937. https://doi.org/10.1016/J.JCLEPRO.2021.127937

Ahmad, S., et al. (2019). *A value tree for multi-criteria evaluation of sustainable aviation fuels*. In International Conference on applied energy [Preprint]. https://www.researchgate.net/profile/Bing-Xu-25/publication/335728465_A_Value_Tree_For_Multi-Criteria_Evaluation_of_Sustainable_Aviation_Fuels/links/5d7a5060a6fdcca980e1c09c/A--Value-Tree-For-Multi-Criteria-Evaluation-of-Sustainable-Aviation-Fuels.pdf. Accessed 11 April 2023.

Anderson, B. J., et al. (2022). Social Science applications in sustainable aviation biofuels research: Opportunities, challenges, and advancements. *Frontiers in Energy Research, 9*, 946. https://doi.org/10.3389/FENRG.2021.771849/BIBTEX

Bergero, C., et al. (2023). Pathways to net-zero emissions from aviation. *Nature Sustainability, 2023*, 1–11. https://doi.org/10.1038/s41893-022-01046-9

del Monte, D. M., Cruz, P. L., & Dufour, J. (2022). SAF production from cameline oil hydrotreatment: A technoeconomic assessment of alternative process configurations. *Fuel, 324*, 124602. https://doi.org/10.1016/J.FUEL.2022.124602

Detsios, N., et al. (2023). Recent advances on alternative aviation fuels/pathways: A critical review. *Energies, 16*(4), 1904. https://doi.org/10.3390/EN16041904

Mofijur, M., et al. (2023). Screening of non-edible (second-generation) feedstocks for the production of sustainable aviation fuel. *Fuel, 331*, 125879. https://doi.org/10.1016/J.FUEL.2022.125879

O'malley, J., Pavlenko, N., & Searle, S. (2021) *Estimating sustainable aviation fuel feedstock availability to meet growing European Union demand*. https://theicct.org/sites/default/files/publications/Sustainable-aviation-fuel-feedstock-eu-mar2021.pdf. Accessed 11 April 2023.

Perea-Moreno, A.-J., et al. (2022). Use of sustainable fuels in aviation – A review. *Energies, 15*(7), 2440. https://doi.org/10.3390/EN15072440

Saaty, T. L. (1977). A scaling method for priorities in hierarchical structures. *Journal of Mathematical Psychology, 15*(3). https://doi.org/10.1016/0022-2496(77)90033-5

Saaty, T. L. (1990). How to make a decision: The analytic hierarchy process. *European Journal of Operational Research, 48*(1). https://doi.org/10.1016/0377-2217(90)90057-I

Saaty, T. (1994). *Fundamentals of decision making and priority theory with the AHP* (Vol. 29, Issue 1). RWS Publications.

Saaty, T. L. (2002). Decision making with the analytic hierarchy process. *Scientia Iranica, 9*(3). https://doi.org/10.1504/ijssci.2008.017590

Shahriar, M. F., & Khanal, A. (2022). The current techno-economic, environmental, policy status and perspectives of sustainable aviation fuel (SAF). *Fuel, 325*, 124905. https://doi.org/10.1016/J.FUEL.2022.124905

Tanzil, A. H., et al. (2021). Strategic assessment of sustainable aviation fuel production technologies: Yield improvement and cost reduction opportunities. *Biomass and Bioenergy, 145*, 105942. https://doi.org/10.1016/J.BIOMBIOE.2020.105942

Xu, B., et al. (2022). Sustainable commercial aviation: What determines air travellers' willingness to pay more for sustainable aviation fuel? *Journal of Cleaner Production, 374*, 133990. https://doi.org/10.1016/J.JCLEPRO.2022.133990

Chapter 10
Optimization of Vortex Generators for a Subsonic Aircraft Wing Using Taguchi Method

An-Jui Hsu, Wen-lih Chen, Ming-shi Hu, and Yi-chung Liu

Nomenclature

C_L	Lift coefficient
C_D	Drag coefficient
CFD	Computational fluid dynamicst
D	The distance between two pairs of VGS
D	The distance between the leading edge of two VGs
VG	Vortex generators
X_{VG}	VG installation location
β	Angle between freestream and VG

10.1 Introduction

As an aircraft stalls under high angle of attack, airflow is separated from wing surface, resulting in drastic lift loss and drag increase. Consequently, the performance of the wing is impaired, jeopardizing the safety of the aircraft. As shown in Fig. 10.1, installing vortex generators (VGs) on the upper surface of aircraft wings is a common practice to mitigate this problem. VGs can introduce high-energy vortices into sluggish boundary layer flow to delay flow separation at high angle of attack. Hence,

A.-J. Hsu (✉) · W.-l. Chen
Department of Aeronautics and Astronautics, National Cheng Kung University, Tainan, Taiwan
e-mail: p46101295@gs.ncku.edu.tw; wlchen@mail.ncku.edu.tw

M.-s. Hu
Aerospace Technology Research and Development Center of ROCAF, Taichung, Taiwan
e-mail: someoneauthorc@institute.com

Y.-c. Liu
Aeronautical Systems Research Division, National Chung-Shan Institute of Science and Technology, Taoyuan City, Taiwan

T. H. Karakoç et al. (eds.), *Energy and Sustainable Aviation Fuels Solutions*, Sustainable Aviation, https://doi.org/10.1007/978-3-031-70694-3_10

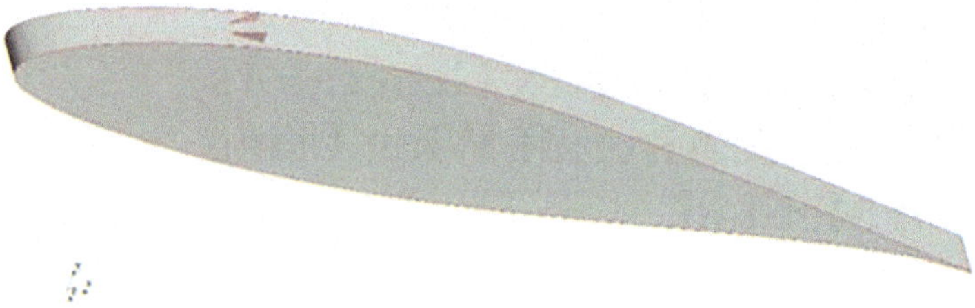

Fig. 10.1 An airfoil fitted with a pair of VGs

they can improve aircraft performance, especially during take-off and landing. In this study, several important parameters of vortex generators are optimized, aiming to find the combinations for producing maximum lift of an aircraft wing at a fixed high angle of attack. Such information can be very useful for the design of vortex generators to improve the safety of an aircraft.

VGs play an important role in improving the aerodynamics characteristics around an aircraft wing. Fin -type VGs are the most popular, and they resemble small vertical wings protruding from the wing surface. They generate tiny wing-tip vortices, which in turn introduce high-energy fluid from free stream into low-momentum boundary layer region. Hence, the latter will better resist adverse pressure gradient and delay flow separation. Some important VG geometrical parameters are length, height, spacing, shape (rectangle, triangle, etc.), and installation position (Lin, 2002). Increasing the height will increase the vortex size and strengthen its intensity, giving rise to higher lift. However, drag will increase as well. Spacing between two adjacent VGs is also influential to the effectiveness of VGs. VGs also come in many different shapes such as rectangle, triangle, or even trapezoid. Each shape has their pros and cons. Among them, triangular VGs produce less drag. The incline angle of a VG, β, is yet another important parameter. An appropriate β angle can further delay flow separation. In general, it is between 10 and 20° (Fouatih et al., 2016). Installing position is very important because VGs are only effective if they are located upstream the separation line (Baldacchino et al., 2018).

10.2 Research Methods

In this study, Taguchi method in conjunction with Computational fluid dynamics (CFD) is adopted to perform parameter optimization. Here, a commercial software STARCCM+ of Siemens is employed for CFD simulation. This is a finite-volume solver. The governing equations are Navier-Stokes equations, and the turbulence model is k-ω SST.

The objective of optimization is maximum life, and the VG factors investigated are β angle, the ratio of the distance between two pairs of VGs and VG height (D/h), the ratio of the distance between the leading edge of two VGs (d/h), and VG installation location (X_{VG}/c). In this study, the height of VG is fixed at 16 mm, which is

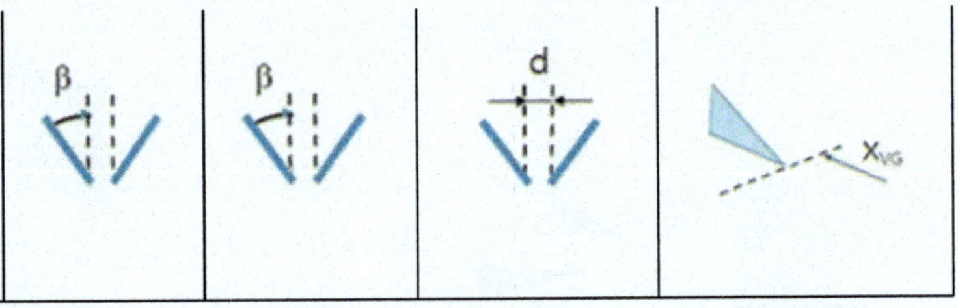

Factor	B (°)	D/h	d/h	X_{VG} /c(%)
	10	8	1	20
Level	15	11	1.5	25
	20	14	2	30

Fig. 10.2 VG factors and their levels for Taguchi method

Table 10.1 The L_9 table for Taguchi optimization

Case	β	D/h	d/h	X_{VG}/c
1	10	8	1	20
2	10	11	1.5	25
3	10	14	2	30
4	15	8	1.5	30
5	15	11	2	20
6	15	14	1	25
7	20	8	2	25
8	20	11	1	30
9	20	14	1.5	20

about the height of boundary layer thickness at 30% chord on the upper surface of the wing.

Figure 10.2 illustrates the VG factors and the levels of factors. The parameter values for Taguchi method are β angle: 10°, 15°, 20°; D/h: 8, 11, 14; d/h: 1, 1.5, 2; X_{VG}/c: 20%, 25%, 30%. Table 10.1 shows the L_9 table for Taguchi optimization.

10.3 Code Validation

Data from NACA63415 airfoil from Bak's experiment (Bak et al., 2000) were used for code validation. Figure 10.3 shows the VG geometry and the computational domain, which has dimensions 20–30 times the chord length in all directions.

Adaptive-mesh-refinement technique has been used to capture complex flow structure. The criterion of mesh refinement is vorticity magnitude, and the levels of refinement are, respectively, set from 1 to 3 for grid independence test. Table 10.2 listed the flow conditions, and the results of the grid independence test are shown in Fig. 10.4. In addition, the comparison of C_L and C_D values is listed in Table 10.3.

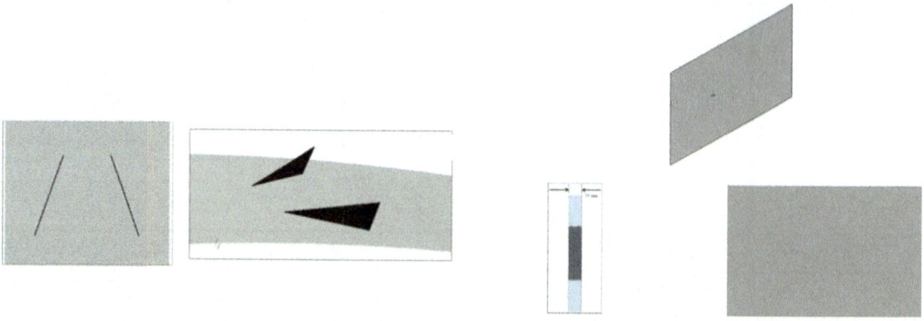

Fig. 10.3 VG geometry and computational domain

Table 10.2 Flow conditions of grid independence test

Feature	Value
Free stream velocity	40 m/s
Density	1.777 kg/m^3
Temperature	300 K
Angle of attack	12°

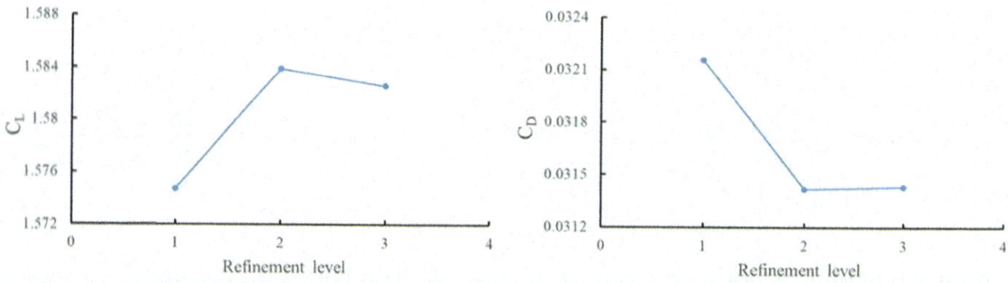

Fig. 10.4 Results of grid independence test

Table 10.3 The values of C_L and C_D, respectively, obtained by three mesh -refinement levels

Refinement level	1	2	3
CL	1.5747	1.5839	1.5826
CD	0.03216	0.03142	0.03143

From Fig. 10.4, the results of lift and drag coefficients returned by two- and three-level refinement are very close. The data in Table 10.3 indicate that their differences are within 1%. This suggests that two-level refinement is good enough to ensure grid-independent solutions.

Figure 10.5 compares experimental and computational C_L and C_D under different angles of attack for the NACA63415 airfoil fitted with VGs. From Fig. 10.5a, b, the agreement between experimental and computational results is excellent before stall angle, which is around 17°. The simulation overestimates C_L and C_D after the stall angle. Nevertheless, Fig. 10.5c suggests that C_L/C_D are well captured in general. To this end, it can be concluded that the present CFD approach can return satisfactory results for this kind of flow.

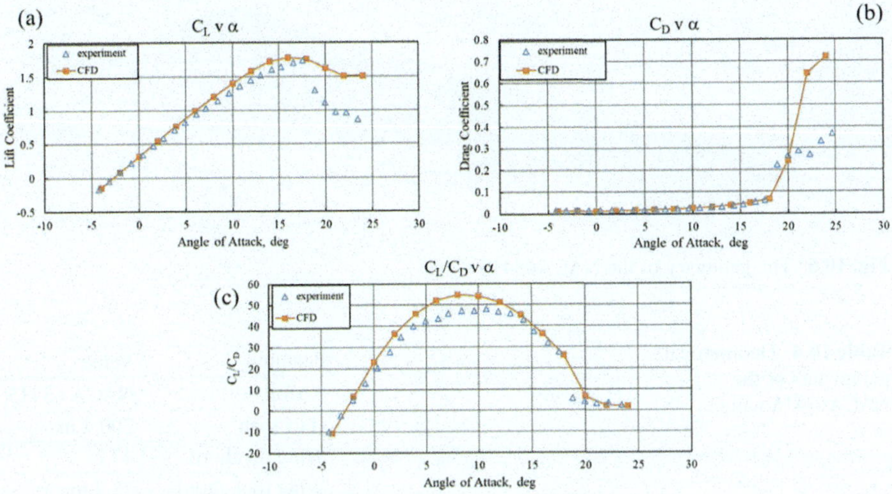

Fig. 10.5 Comparison of experimental and computational C_L and C_D for NACA63415 fitted with VGs

10.4 Results and Discussion

Figure 10.6 shows the geometry of a NACA63415 wing that is investigated in this study. Due to geometrical symmetry, only half of the wing is included in simulation. Some important geometrical parameters of this wing are listed in Table 10.4. Figure 10.7 depicts the installation of VGs on the wing surface according to the L_9 table.

The computational mesh has been refined near wing tip and along the trailing edge. Two levels of adaptive mesh refinement are also employed to better resolve complex flow features. The flow conditions are listed in Table 10.5.

We first simulate the wing flow without VGs to get data for comparison with those from the flow fitted with CGs. There are 13 cases, and among them, the angle of attack varies from −4° to 20°. The results in terms of C_L, C_D, and C_L/C_D are shown in Fig. 10.8. In Fig. 10.8a, it is noted that zero lift occurs at 3°, wing stalls at around 18°, and moment C_L reaches the maximum value of 1.48. From Fig. 10.8c, the maximum C_L/C_D is about 22 at an angle of attack of 2°. Figure 10.9 illustrates the surface streamlines on the suction side at three different angles of attack, 4°, 8°, and 20°. At =4°, there is no flow separation on the suction side, and all streamlines are smooth and straight from the leading to the trailing edge. At a stall angle of 18°, there is a large separation region near the trailing edge. At =20°, this separation is further enlarged and extended more than the mid-chord towards the leading edge at some mid-span location (Fig. 10.9c). VGs should be fitted upstream of the separation line to be effective. Hence, these results offer guidelines for choosing the locations of VGs.

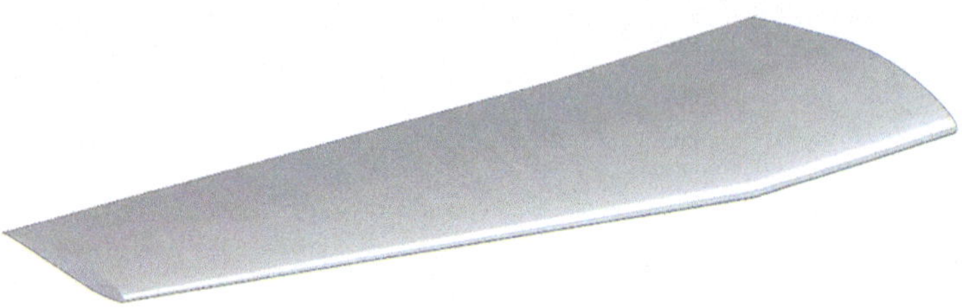

Fig. 10.6 The geometry of the NACA63415 wing

Table 10.4 Geometrical parameters of the NACA63415 wing

Feature	Value
Airfoil	NACA 63415
Half span	200.4 in
Wing tip chord	47.2
Wing root chord	78.7 in
Sweep back angle	3°
Dihedral angle	6°

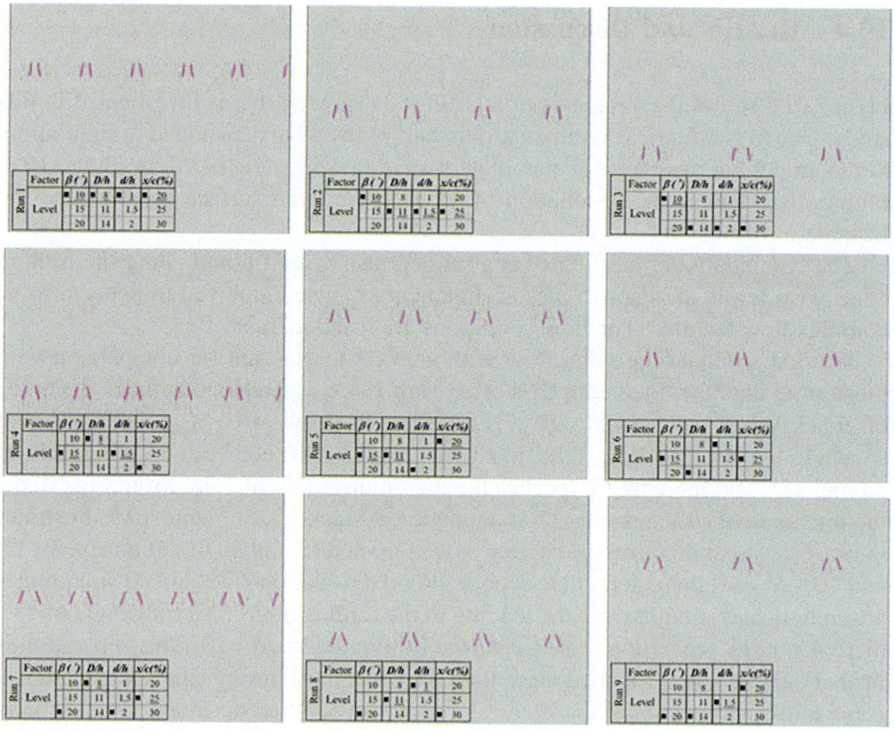

Fig. 10.7 The installation of VGs on the wing surface according to the setting in the L_9 table

Table 10.5 Flow conditions of the NACA63415 wing

Feature	Value
Stream velocity	60 m/s
Density	1.177 kg/m3
Temperature	300 K
Pressure	1 atm

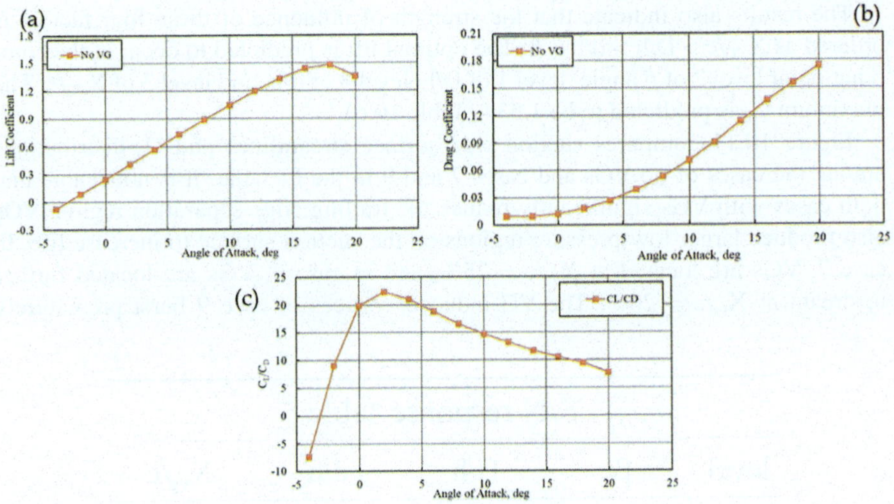

Fig. 10.8 C_L, C_D, and C_L/C_D variations versus angle of attack of the wing without VGs; (**a**) C_L, (**b**) C_D, and (**c**) C_L/C_D

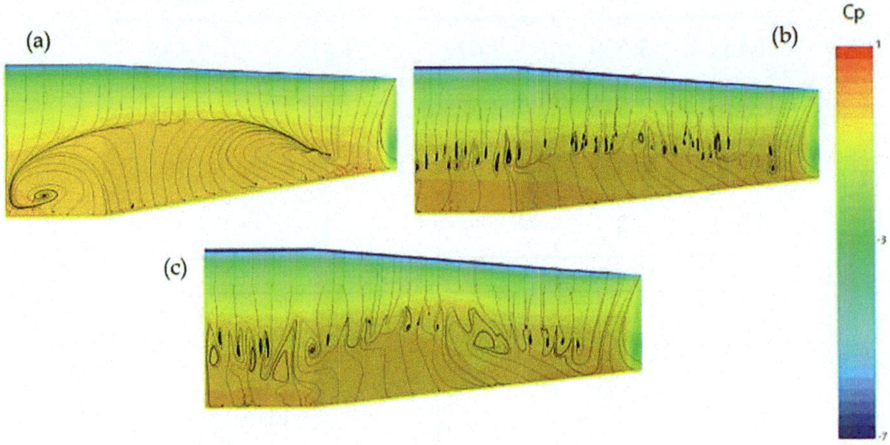

Fig. 10.9 Suction side surface streamlines and Cp distributions at three different angles of attack; (**a**) =4°, (**b**) =18°, and (**c**) =20°

In terms of the Taguchi optimization analysis, the angle of attack and Reynolds number are fixed at 20° and 6.6×10^6, respectively. The objective is to maximize the lift coefficient. Figure 10.10 shows the response graphs of the four factors. Among the four factors, the effect of β angle is not significant, and the optimal value is around 15°. Life increases as (D/h) decreases. On the other hand, life increases as (d/h) increases. X_{VG}/c poses the most significant effect on lift, and the optimal position is on $X_{VG}/c = 0.3$.

The results also indicate that the strength of influence of these four factors is ordered as $X_{VG}/c > D/h > d/h > \beta$. The optimal lift is predicted to occur at the combination of level 2 of β angle, level 1 of D/h, level 3 of d/h, and level 3 of X_{VG}/c. The maximum C_L is predicted to be 1.534 (Table 10.6).

Figure 10.11 compares suction side surface streamlines and C_p distributions among the cases of no VGs and cases 7 and 9 in the L_9 table. It is noticeable that both cases with VGs significantly reduce the trailing edge separation region. VGs also produce larger low-pressure regions on the suction surface to increase life. In case 7, VGs are located at $X_{vg}/c = 25\%$, and in case 9, VGs are located further upstream at $X_{vg}/c = 20\%$. The VG-induced vortices in case 9 burst prematurely

S/N response table

level	β	D/h	d/h	X_{vg}/c
L1	3.548	**3.622**	3.525	3.451
L2	**3.569**	3.529	3.534	3.590
L3	3.559	3.525	**3.617**	**3.635**
Max	3.569	3.622	3.617	3.635
Min	3.548	3.525	3.525	3.451
Diff.	0.021	0.097	0.092	0.184
Rank	4	2	3	1

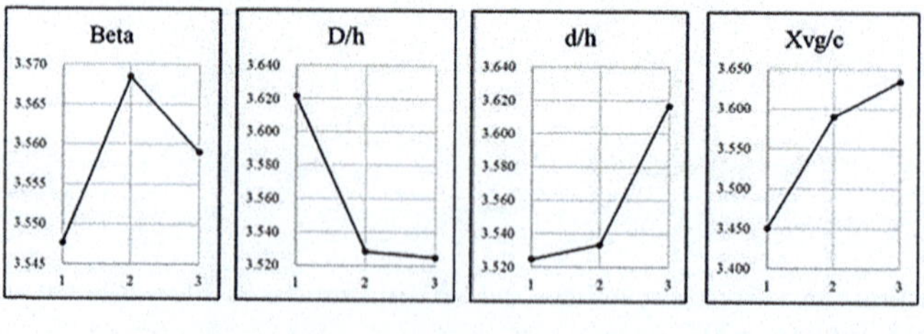

Fig. 10.10 The response graphs of different factors

Table 10.6 The S/N values

Case	β	D/h	d/h	X_{VG}/c	CL(S/N)
1	10	8	1	20	3.470
2	10	11	1.5	25	3.525
3	10	14	2	30	3.649
4	15	8	1.5	30	3.683
5	15	11	2	20	3.490
6	15	14	1	25	3.533
7	20	8	2	25	3.712
8	20	11	1	30	3.572
9	20	14	1.5	20	3.393
Avg.					3.558

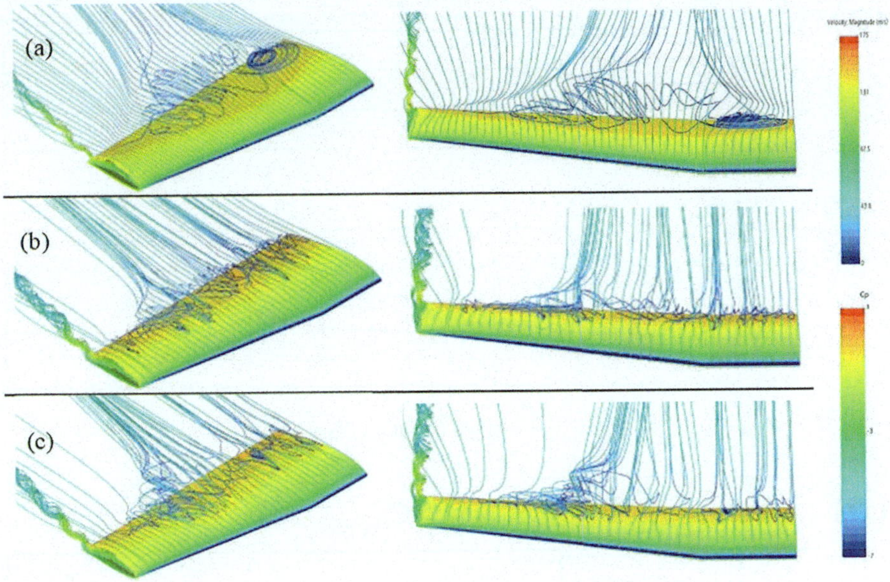

Fig. 10.11 Suction side surface streamlines and C_p distributions; (**a**) no VGs, (**b**) case 7, and (**c**) case 9

upstream of the separation line; hence, their effect on the suppress separation region is less pronounced than case 7. Across the span, VGs near the wing root are more effective than those located near the wing tip. The results manifest the importance of the locations of VGs on improving the aerodynamic performance of the wing.

The maximum C_L acquired by simulation is 1.533, while the maximum C_L predicted by Taguchi method is 1.543. The deviation is only 0.6%, demonstrating the accuracy of the Taguchi optimization method. Figure 10.12 compares C_L and C_D obtained by the wing without VGs and the wing with the optimal VGs at different angles of attack. Figure 10.13 illustrates the percentage of deviation in C_L and C_D

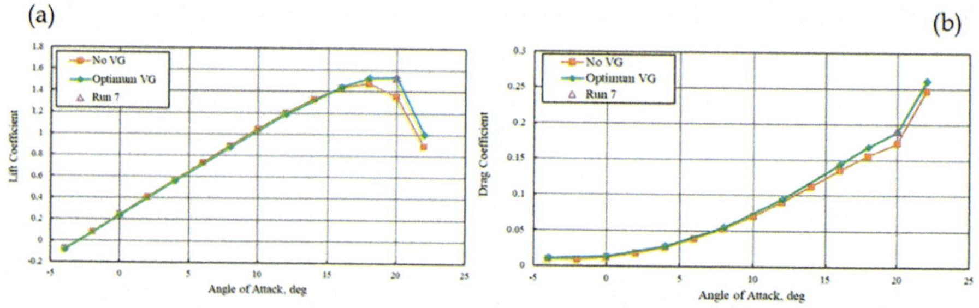

Fig. 10.12 Comparison of aerodynamic performance between wings with and without VGs; (**a**) C_L vs and (**b**) C_D vs

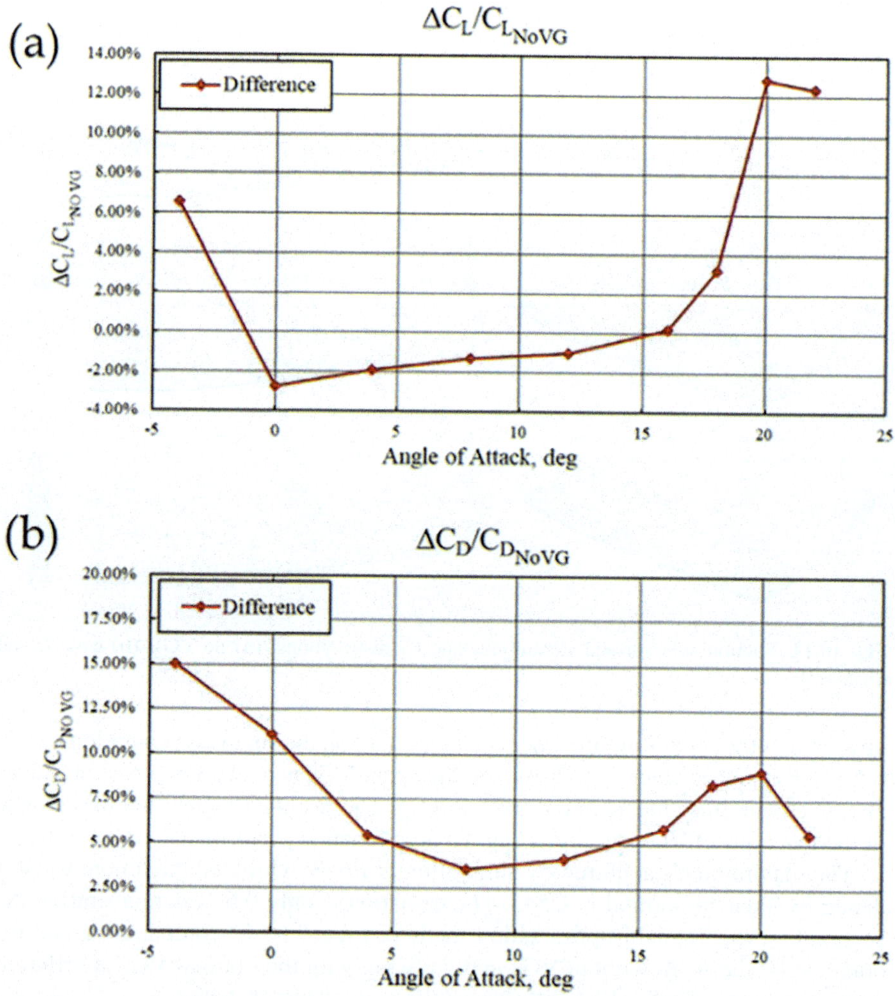

Fig. 10.13 The percentage of deviation in C_L and C_D caused by VGs

caused by VGs. From Fig. 10.12a, the results indicate that C_L returned by both wings are almost identical before the angle of attack reaches 16°. However, beyond 16°, the benefit of VGs begins to stand out. At 20°, the improvement in C_L reaches 12.8% (Fig. 10.13a). As expected, C_D of the wing with VGs is in general higher than that of the wing without VGs (Fig. 10.12b). Nevertheless, Fig. 10.13b shows that the increase in C_D for positive angle of attack is less than 10%.

10.5 Conclusions

Several conclusion remarks can be summarized as follows:

Taguchi method is appropriate for VG optimization analysis.
At lower angle of attack, the presence of VGs poses little effect on lift.
VGs result in a positive effect on increasing lift, especially at an angle of attack near the stall angle.
When the angle of attack is 20°, VGs can improve lift by 12.8%.

In future work, the effect of VG height *h* should be considered because it is also an influential factor in lift.

References

Bak, C., Fuglsang, P., Johansen, J., & Antoniou, I. (2000). *Wind tunnel tests of the NACA 63–415 and a modified NACA 63–415 airfoil*. Risø National Laboratory.

Baldacchino, D., Ferreira, C., Tavernier, D. D., Timmer, W. A., & van Bussel, G. J. W. (2018). Experimental parameter study for passive vortex generators on a 30% thick airfoil. *Wind Energy, 21*(9), 745–765. https://doi.org/10.1002/we.2191

Fouatih, O. M., Medale, M., Imine, O., & Imine, B. (2016). Design optimization of the aerodynamic passive flow control on NACA 4415 airfoil using vortex generators. *European Journal of Mechanics B-Fluids, 56*, 82–96. https://doi.org/10.1016/j.euromechflu.2015.11.006

Lin, J. C. (2002). Review of research on low-profile vortex generators to control boundary layer separation. *Progress in Aerospace Sciences, 38*(4–5), 389–420.

Chapter 11
Airline Technological Services and Airline Passengers' Purchase Intention: An Investigation

Gamze Ozdemir, Fikri Ilgoz, Ibrahim Abdi Mohamed, and Alper Dalkiran

Nomenclature

Pax	Airline passenger
LCC	Low-cost carriers
IATA	International Air Transport Association
ACI	Airport Council International
FFP	Frequent Flyer Program
RFID	Radio Frequency IDentification

11.1 Introduction

The aviation industry has seen quick growth since technological advances. Passenger services and ground-handling activities also grew and evolved to integrate with other services, as demand increased in time (Cam and Durmaz, 2018). Airport and airline operators use technology to improve travelers' journeys (Ağraz, 2006). The airport provides technology-based amenities like digital screens, kiosks, entertainment systems, online ticket purchasing, and mobile apps to support this trend. Airports are improving concession shops and direct sale points through technology to increase customer satisfaction.

Ekici (2012) found that in-flight entertainment systems and plane technologies influence an individual's decision to purchase a ticket (Ekici, 2012). According to the research conducted by Memika and Polat (2023), the classic boarding pass and the online boarding pass were compared, and it was determined that the online boarding pass was more advantageous to spend more efficient time at the airport,

G. Ozdemir · F. Ilgoz · I. A. Mohamed · A. Dalkiran (✉)
Department of Aviation Management, School of Aviation, Suleyman Demirel University, Keciborlu, Türkiye
e-mail: alperdalkiran@sdu.edu.tr

T. H. Karakoç et al. (eds.), *Energy and Sustainable Aviation Fuels Solutions*, Sustainable Aviation, https://doi.org/10.1007/978-3-031-70694-3_11

shortening the time they were in. Efficient airports, which are an important indicator, use more resources to shorten this process and stimulate purchasing behavior (Durmaz and Bahar, 2011). The satisfaction level of higher educated ones and more frequent flying of passengers with children lowers in terms of aviation service perceptions. Passengers with children require quick security and boarding processes at the airport (Kuyucu, 2017; Erten, 2019; Kotro, 2021; Yüksel et al., 2022), as well as extra baggage allowance and price reductions (Güreş et al., 2019). It is also stated that when this process is shortened, it helps to eliminate passenger dissatisfaction and reduce costs. The impact of actively utilizing digital technologies effects the airports positively (Artar and Türkay, 2021; Meydan, 2023) and businesses in airports based on customer purchasing behavior was investigated by Tümer (2022). The current study directs its focus on implementing digital technologies in the purchase process, as opposed to the effect of such technologies (Türkay and Artar, 2021) on purchasing behavior, which is the focus of related research. A survey was conducted with 134 participants. According to research, there are three purchasing processes. In general, a positive relationship was observed in the acceptance and use of digital technologies in purchasing (Beldona et al., 2011; Wehner et al., 2022). It is beneficial to develop digital technologies and to be compatible with processes. A comprehensive literature review has been made, and in the research examined, no study has been encountered that reveals the direct effect of technology on purchasing behavior, and it has been deemed appropriate to examine the ticket purchasing behavior of passengers within the scope of technology.

Data obtained from 200 participants were used in this study. Multiple analyses of the survey data applied in the research were made using the SPSS data analysis application. Accordingly, the subject of this research is to examine the effects of digital technologies, which are used by airport and airline companies and occur in the three stages of the travel process, on the purchasing behavior of passengers. The aim of the research is to reveal the meaningful effects of digital technologies in the ticket purchasing processes of passengers and to provide data to the businesses in the sector and to provide light with papers.

11.2 Method

The Turkiye airline market is growing over than the world's airline market. Therefore, the influence and the correlation between air passengers in Turkiye and the demand for digital technologies of the airline and the airport are crucial for airline marketing. For this reason, a survey was conducted in which 210 passengers participated, and the data of some participants were not included in the analysis in order to provide qualified data. The survey has designed to collect the required data in two stages. Airline passengers who are intent on buying tickets diverted to the first stage. The others have diverted to the second stage after who are not planning to buying tickets in short-term. The question set can be seen in Fig. 11.1.

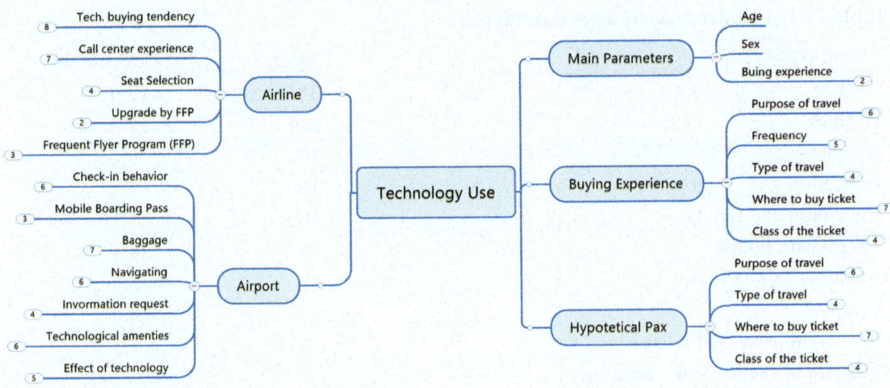

Fig. 11.1 The question set for the research

Passengers who have not traveled before were asked to give hypothetical answers, and explanations were made in order to understand their tendencies toward the airline and airport technologies in each question. Multiple-choice and single-choice research questions were included, separate question titles entered the values of the multiple-choice questions, and these titles were linked to each other with multiple-choice during the analysis.

The analysis application used in the research is SPSS. The research has shaped to prove the hypotheses related to "using digital screens", "analyzing technology use in check-in process", "creating virtual boarding passes", "finding directions at the airport", or "buying tickets electronically" subjects. As a result of the analysis, a statistically and qualitatively meaningful comment will be made on the following pages and suggestions will be made to airline companies and airport operators.

11.3 Results and Discussions

The study would like to find the answers to the below hypothesis. Five hypotheses were determined to understand correlation with the pax intention to buy technology. Those are:

- Which ways do the participants choose for traveling?
- How frequently they are choosing to fly for their travels?
- In which channels they are choosing to buy their tickets?
- Which technologies they would like to use at airports?
- What are their preferences in flight?

In those answers, we have focused on the gender and age distribution first. We have found that there is no significant difference in the participant genders; 48.7% of the participants are women, and 51.3% of the participants are men. Also, the average age of the participants was found as 28 years.

Table 11.1 Frequency analysis of travel types

Travel types	Frequency	Percent (%)
Family	35	17.9
Holiday	22	11.3
Business	19	9.7
Education	18	9.2
Education and family	10	5.1
Family and holiday	8	4.1
Business and holiday	7	3.6
Leisure	7	3.6
Family, holiday, and leisure	6	3.1
Education, family, and education	6	3.1
Business, education, family, and holiday	5	2.6
Sport	2	1.0
Health	1	0.5
Other	49	25.1
Total	195	100.0

Table 11.2 Frequency analysis of travel types

Travel frequencies	Frequency	Percent (%)
Weekly or more	1	0.5
Twice in a month	2	1.0
Monthly	10	5.1
Twice in a year	55	28.2
Yearly	98	50.3
None	29	14.9
Total	195	100.0

The first hypothesis found family visit as the most preferred travel type. Also, education and holiday purposes have been selected with family visit together in participants' answers. So the dominant answer is family visit. The second most preferred travel type is holiday travel. It is seen that 11.3% of the answers point that the respondents likely choose air travel for going on vacation from Table 11.1.

On the other hand, half of the participants selected the yearly travel option, following twice in a year option, 50.3% and 28.2% respectively. Table 11.2 shows the remaining options' frequencies.

We find there is an intensive indication to buying tickets online from the answers of the participants. The Internet is the biggest source of reaching airline tickets, which can be seen in Table 11.3.

Table 11.4 represents the technology using the preferences of the participants. Almost half of the participants do not choose any technology in airports, and the respondents generally use technological solutions over baggage services.

Table 11.3 Ticket-buying preference of the participants

Ticket-buying preference	Frequency	Percent (%)
Airline internet	103	52.8
Airline internet and travel agent internet	24	12.3
Travel agent internet	20	10.3
Airline internet, company buys the ticket	7	3.6
Airline ticket office face to face	5	2.6
Airline internet, airline ticket office face to face	4	2.1
Airline internet, travel agent internet, and airline ticket office face to face	4	2.1
Company buys the ticket	2	1.0
Other	26	13.3
Total	195	100.0

Table 11.4 Technology use preference in the airports

Technology use in airports	Frequency	Percent (%)
No technology use	86	44.1
Self-service baggage drop	31	15.9
Self-service baggage deposit	25	12.8
Mobile baggage cart payment	16	8.2
Early bag drop	10	5.1
Self-baggage wraps	6	3.1
RFID (Radio Frequency IDentification) tag	3	1.5
Other	18	9.2
Total	195	100.0

Table 11.5 Technology use preference in flight

Technology use in flight	Frequency	Percent (%)
None	63	32.3
Both in-flight multimedia and Wi-fi	63	32.3
In-flight Wi-fi	52	26.7
In flight, multimedia service	17	8.7
Total	195	100.0

Table 11.5 represents the technology use preferences in the flight conditions. In total, 32.3% of the participants either did not prefer any technological solution in flight or selected all solutions. If their choice was only one solution, 52 participants expressed they will demand Wi-Fi solution.

11.4 Conclusions

We have introduced a conceptual framework that possesses a fitting level of predictive power when characterizing the buying intention of airline customers to utilize more technological services while traveling. The ability of the framework to forecast passengers' actual behavior was the origin of this power. Furthermore, we have examined the significance of airport behaviors in this framework, and we have effectively integrated this theory into the behavioral intention framework utilized in the airline sector.

The management of an airline must grapple with the problems presented by more affluent and sophisticated passengers, in addition to significant competition from alternatives such as low-cost carriers and high-speed trains, particularly in the aftermath of the Covid-19 pandemic. The study results contribute toward a better understanding of technological product or tariff management from an airline standpoint, providing a competitive edge in this industry. The subject matter under consideration is the potential existence of a multifaceted mechanism governing the generation of airline customers' intention to make repeat purchases or recommendations. This research may be valuable in enlightening airline researchers and operators about the particular functions that in-flight product performance and technology usage serve in promoting customers' favorable post-purchase behavior, which ultimately leads to an expansion in market share (Alamdari, 1999).

The primary factor that significantly influenced the behavioral intention of airline passengers was the performance of the in-flight products, which included the first-order dimensions. The present findings augment the existing knowledge base by providing insights into the utilization of a second-order methodology that recognizes and assesses the performance of a product during its operation. In addition, the parsimonious higher-order product performance typology proposes that researchers and operators formulate theories for complex structures in a more concise manner.

Acknowledgments This research has been funded by the Research, Development and Project Directorate of Suleyman Demirel University under the project code SLP-2022-8932.

References

Ağraz, S. (2006). Havayolu İşletmelerinin İstihdama Katkısı, T.C İstanbul Üniversitesi Sosyal Bilimler Enstitüsü Çalışma Ekonomisi ve Endüstri İlişkileri Anabilim Dalı Doktora Tezi.

Alamdari F. (1999, October). Airline in-flight entertainment: The passengers' perspective. Journal of Air Transport Management, 5(4), 203–209.

Artar, O., & Türkay, U. İ. (2021). Havacılık sektöründe havalimanlarının dijital dönüşümü. *İstanbul Ticaret Üniversitesi Dış Ticaret Enstitüsü Working Paper Series Dergisi, 2*(1), 86–97.

Ayodeji, Y. Rjoub, H., & Özgit, H. (2023). Havalimanlarında sürdürülebilir müşteri bağlılığının sağlanması: Bekleme süresi memnuniyeti ve self-servis teknolojilerinin rolü.

Beldona, S., Racherla, P., & Mundhra, G. D. (2011). To buy or not to buy: Indian consumers' choice of online versus offline channels for air travel purchase. *Journal of Hospitality Marketing & Management, 20*(8), 831–854.

Cam, A. C., & Durmaz, V. (2018). Dijital Havacilik: Güncel Uygulamalarla Gelecekteki Yolcu Deneyimleri. *Electronic Turkish Studies, 13*(26), 251–256.

Canöz, N. (2018). Yabancıların Havayolu Ulaşımını Tercih Etme Nedenleri: Türkiye'ye Ait Havayolu İşletmeleri Üzerine Bir Araştırma. *Selçuk İletişim, 11*(2), 196–213.

Durmaz, Y., & Bahar, R. (2011). Tüketicilerin Satın Alma Davranışları Üzerinde Sosyolojik Faktörlerin Etkisinin İncelenmesine Yönelik Ampirik Bir Çalışma. *Elektronik Sosyal Bilimler Dergisi, 10*(37), 60–77.

Ekici, N. (2012). *Deneyimsel pazarlama ve seyahat deneyimi: Türk Hava Yolları örneği* (Master's thesis, Sosyal Bilimler Enstitüsü).

Erten, P. (2019). Z kuşağının dijital teknolojiye yönelik tutumları. *Gümüşhane Üniversitesi Sosyal Bilimler Enstitüsü Elektronik Dergisi, 10*(1), 190–202.

Güreş, N., Arslan, S., Yüksel, C., & Yilmaz, H. (2019). Çocuklu Yolcuların Havacılık İşletmelerine Yönelik Beklenti ve Memnuniyetlerinin Araştırılması. *Ömer Halisdemir Üniversitesi İktisadi ve İdari Bilimler Fakültesi Dergisi, 12*(1), 120–132.

Kotro, A. (2021). X, Y Ve Z Kuşağı Havayolu Tüketicisinin Satın Alma Davranışı, T.C. Kocaeli Üniversitesi Sosyal Bilimler Enstitüsü Havacılık Yönetimi Anabilim Dalı Havacılık Yönetimi Bilim Dalı Yüksek Lisans Tezi.

Kuyucu, M. (2017). Y Kuşağı ve Teknoloji: Y Kuşağının İletişim Teknolojilerini Kullanım Alışkanlıkları. *Gümüşhane Üniversitesi İletişim Fakültesi Elektronik Dergisi, 5*(2), 845–872.

Memika, T., & Polat, T. K. (2023). Internet of Things supported airport boarding system and evaluation with fuzzy. *Intelligent Automation & Soft Computing, 35*(3).

Meydan, C. H. (2023). Havayolu işletmelerinde dijital dönüşüm uygulamaları üzerine bir inceleme. *Journal of Aviation Research, 5*(1), 65–82.

Tümer, S. (2022). Endüstriyel Satın Almada Dijital Teknolojilerin Kabulü: TAM ve TOE Modelleri Kapsamında Havacılık Sektöründe Bir Uygulama.

Türkay, U., & Artar, O. (2021). *Havacılık Sektöründe Havalimanlarının Dijital Dönüşümü* (Working Paper Series, Vol. 2, Issue 1, pp. 86–97).

Wehner, C., López-Bonilla, L. M., & López-Bonilla, J. M. (2022). Consumer behaviour in air transportation: Comparison between Spanish and German passengers. *Travel Behaviour and Society, 28*, 264–272.

Yüksel, F. Ş., Kalan, O., & Melek, I. Ş. I. K. (2022). Havaalanlarında Dijital Dönüşüm Risklerinin Değerlendirilmesi. *Çukurova Üniversitesi Mühendislik Fakültesi Dergisi, 37*(3), 781–792.

Chapter 12
The Effect of Air-Blast Injector Design on Swirl Number and Spray

Yu-Ta Chen, Tony Yuan, Shih-Tu Wu, and Chun-Hsiang Liao

12.1 Introduction

Air-blast injectors atomize spray by the velocity difference between liquid spray and airflow to transfer kinetic energy. Liquid film breaks up into tiny droplets rapidly due to centrifugal force and the effect of air vortex. Axial guide vane is commonly used to generate air vortex, and tangential entry structure is commonly used to generate liquid vortex (Lefebvre, 1989; Stiesch, 2003; Chen, 2020).

A simplified air-blast injector referring to the fuel nozzle of TFE1042–70 was adopted in the research. Various design parameters were used to discuss the influence of air-blast injector design on the intensity of air vortex and the spray characters. Swirl number is a quantitative indicator of air vortex intensity in this research.

12.2 Experimental Methods

Design parameters in this research are shown in Fig. 12.1. The internal liquid vortex injector of the air-blast injector utilizes a vortex disc with a two-injection hole design and a length-to-diameter ratio of 4.2 (L = 2.5 mm, D = 0.6 mm) at the outlet section. The external gas flow channel generates vortex airflow through axial guiding blades. In the study, various parameter settings were employed for the axial guiding blades and gas flow channel.

In the study, the Malvern Spraytec-STP2000 particle size analyser was used to measure the average particle size of the spray. The Sauter mean diameter (SMD) was used to analyse the average particle size of the spray.

Y.-T. Chen (✉) · T. Yuan · S.-T. Wu · C.-H. Liao
Department of Aeronautics and Astronautics, National Cheng Kung University, Tainan, Taiwan

T. H. Karakoç et al. (eds.), *Energy and Sustainable Aviation Fuels Solutions*, Sustainable Aviation, https://doi.org/10.1007/978-3-031-70694-3_12

MODEL	S45	S90	M15	M30	M45
Width of airflow channel	1.5	1.5	2.5	2.5	2.5
Angle of vane	45°	90°	15°	30°	45°
Number of vane	8	4	12	12	8
MODEL	M45+	M90	M90a	L45	L90
Width of airflow channel	2.5	2.5	2.5	3.5	3.5
Angle of vane	45°	90°	90°	45°	90°
Number of vane	12	4	8 (airfoil	8	4

Fig. 12.1 Specifications of air-blast injector designs

$$SMD = \sum_i \frac{n_i d_i^3}{n_i d_i^2}$$

where d_i represents the median particle diameter in each diameter interval and n_i represents the number of particles corresponding to each diameter interval. It represents the ratio of droplet volume to surface area.

In the present study, the planar laser-induced fluorescence (PLIF) technique is employed to perform detailed observations of individual spray structure in a multi-spray setting (Jung et al., 2000; Notaro et al., 2019; Yuan et al., 2005, 2006). The researchers conducted a comparison between the hot-fire and cold-flow observations of nitrogen tetroxide/monomethylhydrazine (NTO/MMH) impinging spray. By analysing the mass distribution in the cold-flow experiment using simulants, they were able to predict the temperature distribution of the impinging combustion of NTO/MMH. In this study, the PLIF technique was used to semi-quantitatively describe the distribution of the spray.

A 3D model of the air-blast injector was created using Solidworks, and flow field simulation was performed using Simcenter STAR-CCM+. In the flow field simulation, both steady-state and transient simulations were conducted. For the steady-state simulation, the Reynolds-average Navier-Stokes (RANS) turbulence model and k-epsilon turbulence model were employed to calculate the average vortex number of the flow field. For the transient simulation, dynamic grid refinement was applied at the gas-liquid interface to capture liquid fragmentation, and the large eddy simulation (LES) turbulence model was used. The two-phase flow was simulated using the Eulerian multiphase approach with the volume of fluid (VOF) model, taking into account the effects of surface tension and viscosity on the interaction between the fluids.

12.3 Results and Discussion

Computational Fluid Dyanmics (CFD) simulation shows that the liquid spray which did not interact with air vortex spread out in petal shape, developed into ligaments, and finally broke into droplets without a hollow region. The liquid spray which

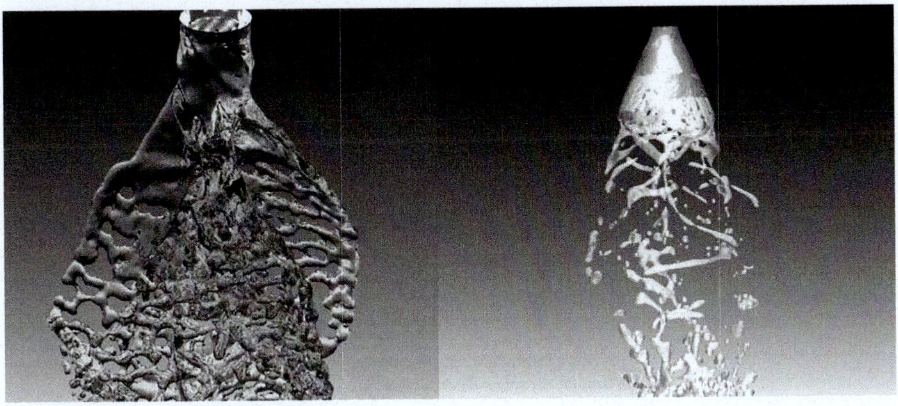

Fig. 12.2 The results of CFD simulation

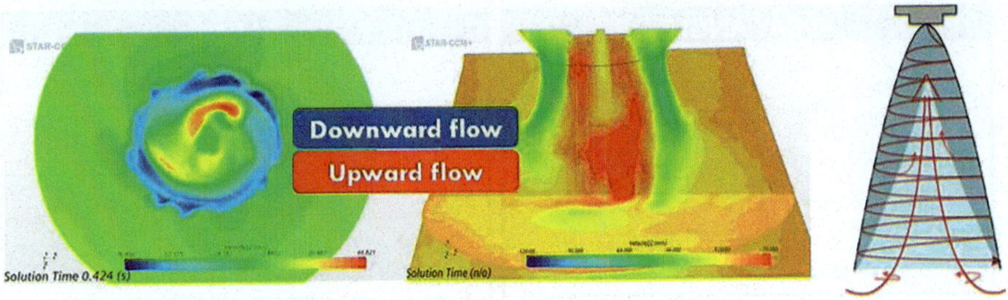

Fig. 12.3 The trajectory of airflow from CFD simulation

interacted with air vortex speeded up the atomization process. The air vortex rotated counterclockwise and formed a cone-shaped air vortex layer with a low-pressure region inside under centrifugal force (Fig. 12.2).

The downstream airflow moved upwards into a cone-shaped air vortex layer in counterclockwise direction to make the spray angle expand and form a hollow region. The external airflow flowed inwards to the low-pressure region inside the cone-shaped air vortex layer through the gaps on the spray surface (Fig. 12.3).

The hollow phenomenon of spray includes the size of the hollow region and the intensity of the hollow phenomenon. The size of the hollow region is defined as the distance between two maximum poles in 2D droplet mass probability scatter diagram. The hollow intensity is defined as the distance between the maximum and the minimum poles in 2D droplet mass probability scatter diagram (Fig. 12.4).

With a larger downstream distance from the injector, the spray angle was decreased under gravity. The SMD, the non-uniformity, and the hollow phenomenon of the spray decreased as the liquid spray continuously interacted with the surrounding air (Fig. 12.5).

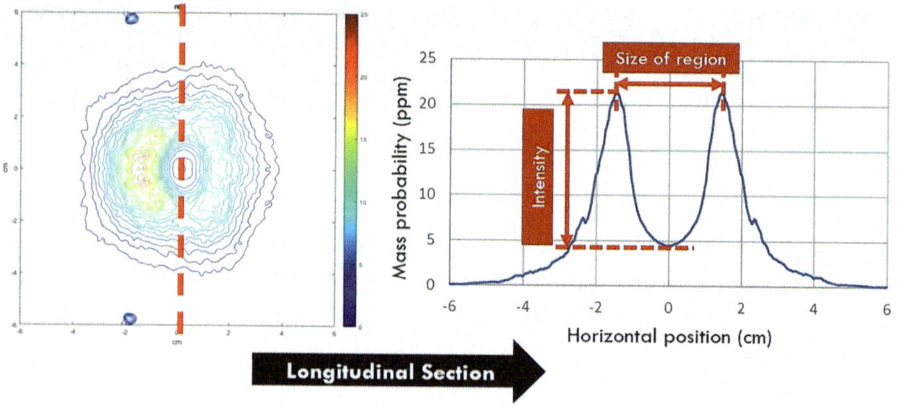

Fig. 12.4 The definition of the hollow phenomenon

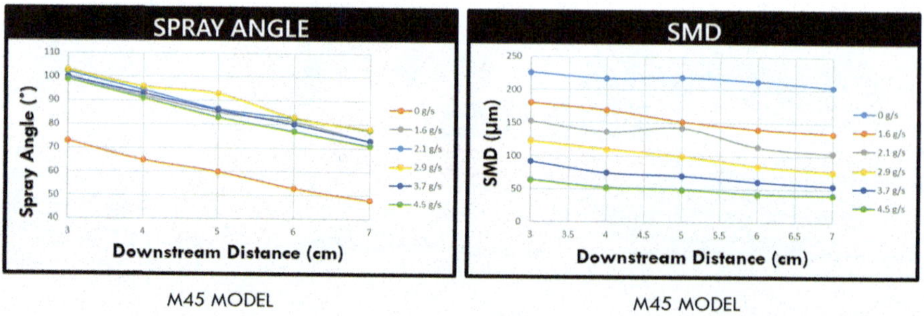

Fig. 12.5 The development of spray

As the air mass flow rate kept increasing, the axial velocity of air increased greater than the tangential velocity. The air vortex swirl number increased first and then decreased with the increasing air mass flow rate (Fig. 12.6).

Take the M45 model, for example, the air mass flow rate increased from 0 g/s to 1.6 g/s, the swirl number of air vortex and the spray angle increased, and the hollow phenomenon was formed. As the air mass flow rate kept increasing to 4.5 g/s, the swirl number of air vortex and the spray angle decreased. The airflow field had a larger axial velocity, a larger recirculation zone, and a larger recirculation velocity; with the continuously growing air mass flow rate, the spray compressed and concentrated, the spray angle reduced slightly, the hollow region and the hollow intensity enhanced (Fig. 12.7), and the SMD of spray became smaller (Fig. 12.8).

The angled guide vane made the airflow steer and increased the tangential velocity of airflow, and the swirl number of air vortex increased (Fig. 12.9).

With larger swirl number of air vortex, the spray angle was larger, the hollow region and the hollow intensity were weaker, the uniformity of spray was larger, and the SMD of spray was smaller (Fig. 12.10).

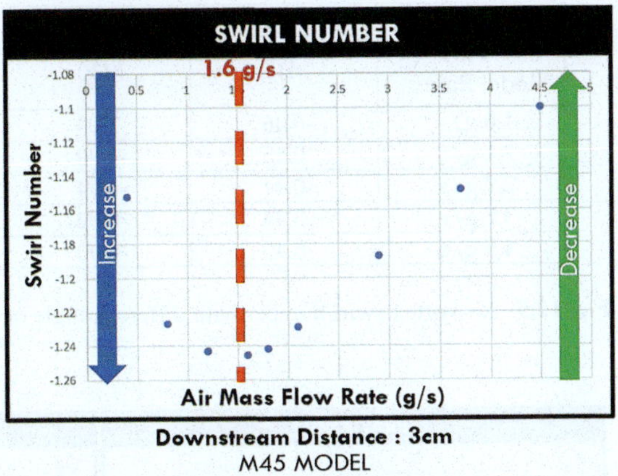

Fig. 12.6 Swirl number of air vortex with air mass flow rate

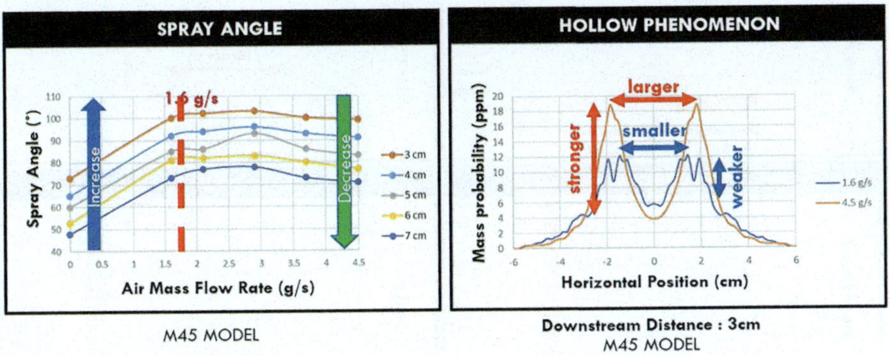

Fig. 12.7 Characteristics of spray with air mass flow

SMD (μm)					
Mass flow rate \ Distance	3cm	4cm	5cm	6cm	7cm
0 g/s	226.1	216.8	218.4	212.3	203.4
1.6 g/s	180.3	169.3	152.0	140.6	134.2
2.1 g/s	152.5	136.6	142.2	113.7	104.0
2.9 g/s	123.3	111.0	100.0	85.3	76.9
3.7 g/s	91.8	74.9	70.2	61.4	55.4
4.5 g/s	64.0	52.6	49.8	43.1	41.6
(Increase ↓)	Decrease ↓	Decrease ↓	Decrease ↓	Decrease ↓	Decrease ↓

Fig. 12.8 SMD of spray with air mass flow rate

The design with greater number of guide vanes had greater axial velocity of airflow, and the swirl number of air vortex was smaller. The spray angle was smaller, the hollow region and the hollow intensity were stronger, and the uniformity of spray decreased (Fig. 12.11).

Mass Flow \ Model Rate	Swirl Number			
	M15	M30	M45	M90
1.6 g/s	-0.46	-1.03	-1.25	-1.44
2.1 g/s	-0.45	-1.00	-1.23	-1.47
2.9 g/s	-0.44	-0.94	-1.19	-1.44
3.7 g/s	-0.43	-0.92	-1.15	-1.42
4.5 g/s	-0.49	-0.90	-1.10	-1.39

Fig. 12.9 Air vortex swirl number with different angle of guide vanes

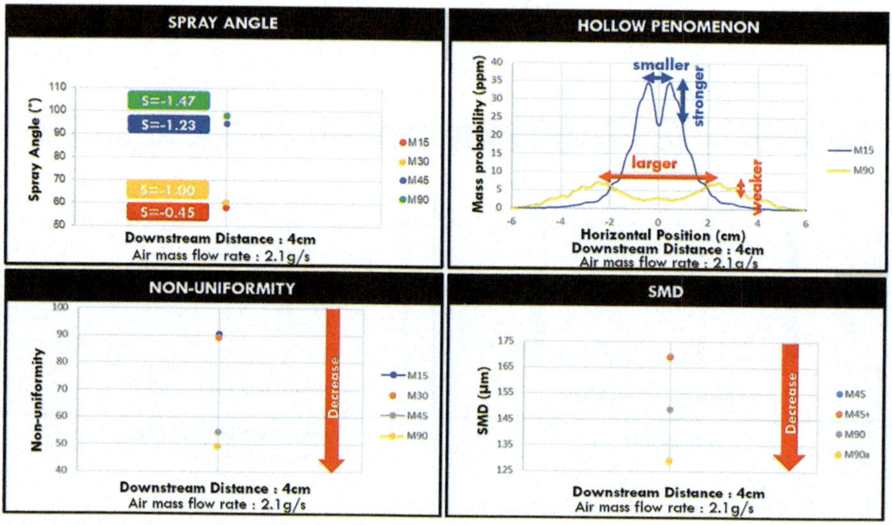

Fig. 12.10 Characteristics of spray with guide vane design

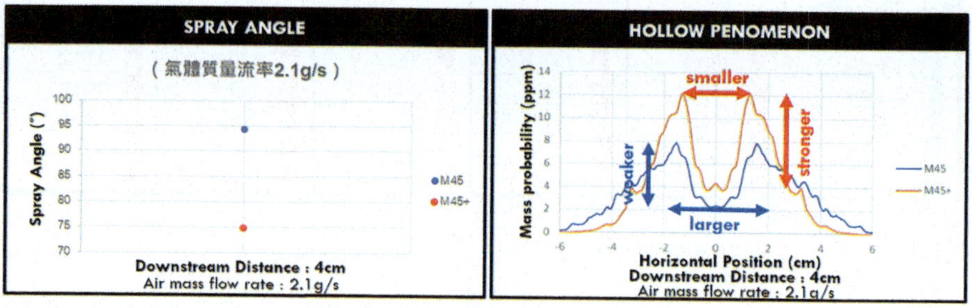

Fig. 12.11 Characteristics of spray with the number of vanes

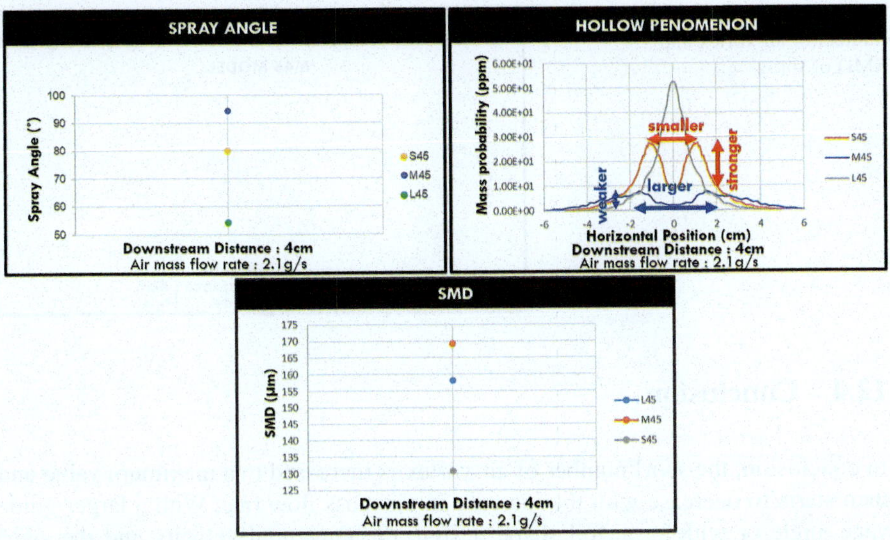

Fig. 12.12 Characteristics of spray with the width of airflow channel

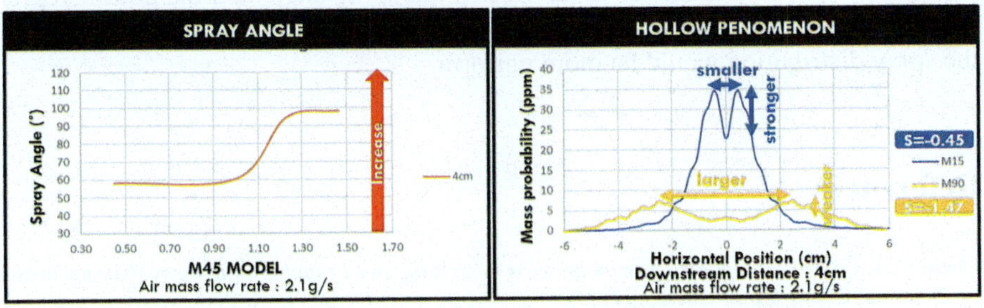

Fig. 12.13 The swirl number of air vortex and spray characteristics

With wide airflow channel, the spray angle would not expand completely due to slow airflow. With narrow airflow channel, the spray was compressed by airflow with large axial velocity, the spray angle and the SMD was small, and the hollow region and the hollow intensity were larger (Fig. 12.12).

To sum up the relation between the swirl number of air vortex and spray, the spray angle had a positive correlation with the swirl number of air vortex, and the hollow region and the hollow intensity had a negative correlation with the swirl number of air vortex (Fig. 12.13).

With the increase in air vortex, swirl number can also decrease the SMD of spray (Fig. 12.14).

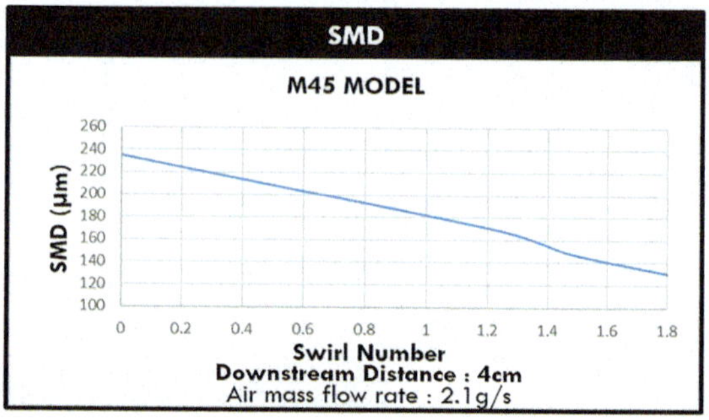

Fig. 12.14 The swirl number of air vortex and SMD of spray

12.4 Conclusion

In conclusion, the swirl number of air vortex is increased to a maximum value and then starts to decrease with the increase in air mass flow rate. With a larger guide vane angle or with an airfoil shape design, the tangential velocity and the swirl number of air vortex would increase. The design with a greater number of vanes made the axial velocity of airflow larger, and the swirl number of air decreased.

With a larger swirl number of air vortex, the spray angle and the hollow region would be larger, the hollow intensity and the SMD of spray would be smaller, and the spray distribution would be more uniform.

References

Chen, C. J. (2020). *The effect of swirl injector design on swirl number and spray* (Master thesis. National Cheng Kung University).

Jung, K., Yoon, Y., & Hwang, S. S. (2000). *Spray characteristics of impinging jet injectors using imaging techniques*. AIAA Paper 2000–3396.

Lefebvre, A. H. (1989). *Atomization and sprays*. Hemisphere.

Notaro, V., Khare, P., & Lee, J. G. (2019). Mixing characteristics of non- Newtonian impinging jets at elevated pressures. *Flow, Turbulence and Combustion, 102*, 355–372.

Stiesch, G. (2003). *Modeling engine spray and combustion processes*. Springer.

Yuan, T., Chen, C., & Huang, B. (2005). *A PLIF Observation of the impingements of NTO/MMH simulants for a 5-lbf rocket*. In Proceedings of the international symposium on heat and mass transfer in the spray systems, 2005.

Yuan, T., Chen, C., & Huang, B. (2006). Optical observation for the impingements of nitrogen tetroxide/monomethylhydrazine simulants. *AIAA Journal, 44*(10), 2259–2266.

Chapter 13
Properties and Specifications of Sustainable Aviation Fuels and Conventional Aviation Fuels

Mhd Bashar Al Kazzaz and Beneda Károly

Nomenclature

ASTM	American Society for Testing Materials
ATJ	Alcohol to jet
CAF	Conventional aviation fuel
CHJ	Catalytic hydrothermolysis jet
EASA	European Union Aviation Safety Agency
FT	Fischer-Tropsch
FT-SKA	Fischer-Tropsch synthetic kerosene with aromatics
HC-HEFA	Hydrocarbon-hydroprocesed esters and fatty acids
HEFA	Hydroprocessed esters and fatty acids
HH-SPK	Hydroprocessed hydrocarbon-synthesised isoparaffinic kerosene
IATA	International Air Transport Association
ICAO	International Civil Aviation Organization
MSW	Municipal solid waste
SAF	Sustainable aviation fuel
SIP	Synthetic isoparaffins

13.1 Introduction

The sustainable aviation fuel (SAF) properties and specifications are critical for ensuring the safe and efficient operation of the TKT-1 engine's fuel system and significantly impacting its performance.

M. B. Al Kazzaz (✉) · B. Károly
Faculty of Transportation Engineering and Vehicle Engineering, Department of Aeronautics and Naval Architecture, Budapest University of Technology and Economics, Budapest, Hungary
e-mail: balkazzaz@edu.bme.hu; beneda.karoly@kjk.bme.hu

T. H. Karakoç et al. (eds.), *Energy and Sustainable Aviation Fuels Solutions*, Sustainable Aviation, https://doi.org/10.1007/978-3-031-70694-3_13

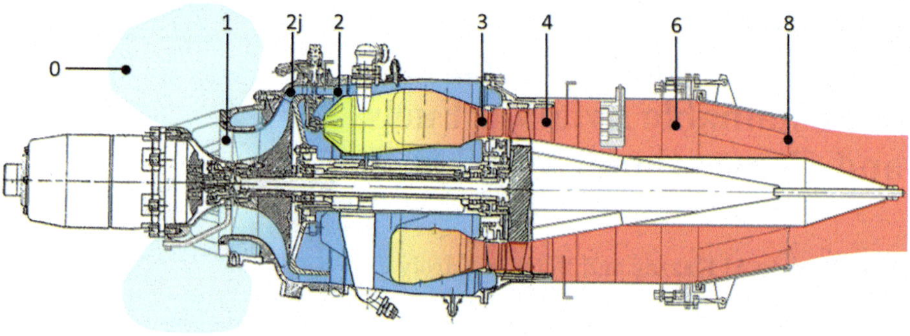

Fig. 13.1 TKT-1 engine

This TKT-1 engine is a single spool engine (Fig. 13.1), which is located on a test bench at the Department of Aeronautics and Naval Architecture of Budapest University of Technology and Economics; it has a centrifugal compressor, axial turbine, and annular combustion chamber which has four nozzles that inject the required fuel. Currently, these four nozzles are sourced by one fuel system. A modification to the fuel system will be done by installing a new additional fuel system to the engine to feed one nozzle with SAF and the remaining three nozzles will stay fed with conventional aviation fuel (CAF).

One of the primary goals of investigating the SAF properties is to ensure the fuel system‘s integrity and dependability. It specifies the design and construction requirements for various components, such as fuel tanks, pumps, filters, valves, and pipelines, to prevent fuel leaks, reduce the risk of fire or explosion, and maintain fuel system reliability. Compliance with the specifications' requirements helps reduce potential hazards and improves the operations and safety of the TKT-1 engine and its components.

Furthermore, the SAF properties and specifications address the efficiency and performance of the fuel system. It specifies parameters such as fuel flow, pressure, temperature, and fuel consumption, all of which have a direct impact on the engine's endurance and overall operational capabilities. Considering the SAF properties and specifications allows us to optimise fuel system design, ensuring optimal fuel utilisation and lowering fuel-related operational costs.

Environmental factors are also taken into account in SAF specifications. It includes production, fuel handling and storage requirements, and measures to reduce fuel emissions and minimize the negative pollution impact.

In summary, the SAF properties and specifications have a significant impact on the design, performance, and safety of the engine's fuel system. It establishes standards and guidelines to ensure system integrity, maximise fuel efficiency, and promote environmental responsibility.

13.2 Chemical Structure and General Properties of CAF

Kerosene is a chemical mixture of alkanes (saturated hydrocarbons) with molecular chain lengths ranging from 10 to 18 carbons and aromatics (Fig. 13.2).

Jet fuels typically have a boiling range of 150–270 °C. This range is somewhere between the boiling points of petrol and diesel used in automobiles (Wilbrand, 2018).

CAFs are made up of a complex mixture of hydrocarbon molecules that can be categorised into four major families (plus trace compounds) (Holladay, 2020):

- n-alkanes 20%,
- Isoalkanes 33%.
- Aromatics ~20%.
- Cycloalkanes 26%.
- Trace compounds.

n-alkanes and isoalkanes: Burn cleanest, are non-toxic, and drive gravimetric energy density,
Aromatics: Form soot and particles, are toxic and cancerogenic, affect sealings and O-rings, and drive volumetric energy content.
Cycloalkanes: Medium clean burning, non-toxic, and balanced energy content.
Trace compounds: Form particles and sulphur, nitrogen, and heavy metals and affect lubricity and conductivity.

Jet A-1 is a kerosene grade of fuel suitable for most turbojet engines which meets the requirements of the ASTM specification D1655 (ASTM, 2022b), IATA Guidance

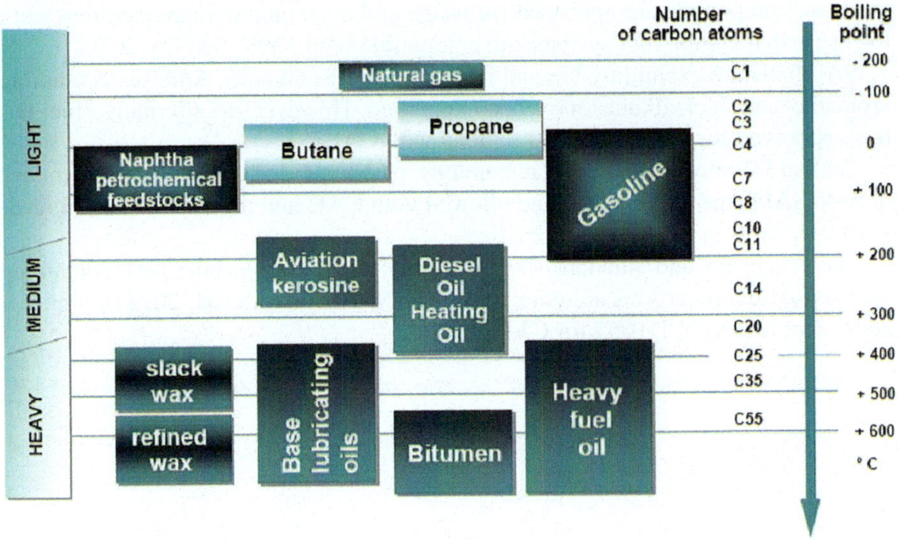

Fig. 13.2 Chain length and boiling temperature of typical hydrocarbon products. (U.Venkata, 2010)

Table 13.1 Properties for Jet A-1 fuel

Property	Max value	Unit
Density at 15 °C	775–840	[kg/m³]
Flash point	38	[°C]
Freezing point	−47	[°C]
Distillation endpoint	300	[°C]
Aromatics content	25	[vol.-%]
Sulphur content	0.3	[mass-%]

Material, and NATO Code F-35. The most important Jet A-1 properties are shown in Table 13.1.

The Jet A-1 typically includes many more parameters, many of which are unique to jet fuel such as thermal oxidative stability. (All other parameters are detailed in ASTM D1655).

13.3 Chemical Structure and General Properties of SAF

One of the most important technical requirements for SAF is to meet or exceed current jet fuel specifications, according to current standards; it is blended with CAF at no more than 50% by volume. The present mix limitations are in place to ensure the appropriate level of safety and compatibility with engine fuel systems.

This is mainly due to the high quantity of aromatic chemicals found in traditional fossil-based fuels, which are required for some systems that utilise nitrile rubber seals. Higher blend limits are anticipated to be allowed in the future as synthetic aromatic compounds are approved for usage and most engine manufacturers work to ensure that engine fuel systems are compatible with 100% SAF by 2030.

SAF fuels are expanding beyond n-alkanes and isoalkanes. Annexes containing aromatics and cycloalkanes have been approved. However, no alternative fuel has been approved for more than a 50% blend with CAF.

Table 13.2 shows the seven technology pathways that can currently produce drop-in SAFs, the maximum blend allowed with CAF, and the most common feedstocks for drop-in SAFs (IATA, 2020).

The properties and standard specifications for aviation turbine fuel–containing synthesised hydrocarbons are detailed in ASTM D7566 (ASTM, 2022a), which is equivalent to ASTM D1655 for CAF.

Table 13.2 Technology pathways for producing SAF

No	Technology	Maximum blending with CAF %	Feedstocks
1	FT and FT-SKA	50	Wastes (MSW, etc.), coal, gas, sawdust
2	HEFA	50	Vegetable oils: palm, camelina, jatropha, used cooking oil
3	HH-SPK or HC-HEFA	10	Oils produced from (*Botryococcus braunii*) algae
4	SIP	10	Sugarcane, sugar beet
5	ATJ	50	Sugarcane, sugar beet, sawdust, lignocellulosic residues (straw)
6	CHJ	50	Waste oils or energy oils

13.4 Conclusion

SAF has similar energy content and combustion characteristics as standard aviation fuels, allowing them to be readily incorporated into existing engine fuel systems. Furthermore, it offers considerable reductions in greenhouse gas emissions, including carbon dioxide, Sox, NOx, soot, and particulate matter contribute to enhanced air quality and climate change mitigation. The above-detailed properties of SAF make it very easy to implement in the already available TKT-1 test bench currently fed with CAF and minimise the need for costly modifications. SAFs' qualities reveal their potential to change the aviation sector into a more sustainable and ecologically friendly future.

References

ASTM. (2022a). *Standard specification for aviation turbine fuel containing synthesized hydrocarbons - D7566.*

ASTM. (2022b). *Standard specification for aviation turbine fuels – D1655.*

Holladay, J. (2020). *Sustainable aviation fuel: Review of technical pathways.* https://www.energy.gov/sites/prod/files/2020/09/f78/beto-sust-aviation-fuel-sep-2020.pdf

IATA. (2020). *Sustainable aviation fuel: Technical certification.* https://www.iata.org/contentassets/d13875e9ed784f75bac90f000760e998/saf-technical-certifications.pdf

U.Venkata, R. (2010). *Crude oil characteristics and refinery products.*

Wilbrand, K. (2018). Potential of fossil kerosene. In *Biokerosene: Status and prospects* (pp. 43–57).

Chapter 14
Energy Minimization in CO_2 Capture in a Natural Gas Power Plant

Rukiye Öztekin and Delia Teresa Sponza

14.1 Introduction

Industrial activities have led to an increasing level of carbon emissions in the atmosphere because of the growing level of industrialization and urbanization in many developing countries (O'Neill & Oppenheimer, 2003). This has also led to a significant increase in the global atmospheric concentration of anthropogenic greenhouse gases, such as CO_2(g), leading to global warming and climate change. According to the International Energy Agency, from 37.1 GT of CO_2(g) produced in 2019, 35 gross tons belong to the energetic sector. The environmental effect of greenhouse gases considers that CO_2(g) contributes 60% of the effects of global warming (Olajire, 2010). The options to reduce total CO_2(g) emissions can be summarized into three options: (1) enhance conversion efficiency, (2) use a low/carbon-free fuel, and (3) improve CO_2(g) capture. For reducing CO_2(g) emissions from power plants, capture and storage (CCS) is considered a promising option. The most common technology to capture CO_2(g) is chemical absorption with amines, especially monoethanolamine (MEA). This process, however, requires a significant amount of energy for solvent regeneration which penalizes the efficiency of the electricity production.

The transported liquefied natural gas (LNG) is regasificated to be burned in the natural gas combined cycle (NGCC) to produce electricity (Lee & Lim, 2019; Lindqvist et al., 2014). However, a significant amount of cold energy is wasted on the seawater during the LNG regasification (Bao et al., 2019; Lee et al., 2022; Bao et al., 2017). Additionally, in NGCC, the hot energy from the condenser, which is used to condense the low-pressure steam, is disposed into seawater

R. Öztekin (✉) · D. T. Sponza
Faculty of Engineering, Department of Environmental Engineering,
Nameof 1st Dokuz Eylül University, İzmir, Türkiye
e-mail: delya.sponza@deu.edu.tr

T. H. Karakoç et al. (eds.), *Energy and Sustainable Aviation Fuels Solutions*,
Sustainable Aviation, https://doi.org/10.1007/978-3-031-70694-3_14

(Paudel & Bandhauer, 2018). Therefore, to increase overall energy efficiency, the utilization of waste cold and hot energy from the LNG regasification process and NGCC, respectively, is crucial to achieving lower fuel usage. It is possible to recover the waste hot and cold energy for CCS (Lim et al., 2021).

In this study, waste LNG cold energy and hot energy from NGCC were operated as simulants to maximize energy efficiency for CCS. In several coolers to lower the temperature of the pure amine solution in the amine cleaning system, the use of LNG cold energy, waste hot energy, and CO_2(g) liquefaction was implemented to improve overall energy efficiency. It is possible to minimize and maximize the absorber and scraper temperature by using waste hot and cold energy; CO_2(g) absorption efficiency and amine regeneration efficiency are also maximized with lower energy consumption. With this study, more economic and environmental protection will be provided in LNG power plants.

14.2 Materials and Methods

The suggested processes are composed of the following four steps: LNG regasification process, NGCC, capture process (CCP), and CO_2(g) liquefaction process (CLP).

14.2.1 LNG Regasification Process

Definition of LNG Regasification Process; It is the process of regasifying the LNG transported from the specific location where the NG resources are located, and the LNG to be burned in the gas-fired power plant through the combined cycle in the gas/steam turbine.

14.2.2 Natural Gas Combined Cycle (NGCC)

The vaporized NG is burned at the combustor in NGCC. In NGCC, steam for power generation is produced at the heat recovery steam generator (HRSG), and flue gas is emitted (Nag & De, 1997).

14.2.3 CO_2(g) Capture Process (CCP)

CO_2(g) in flue gas from the boiler is captured during the CCP. Amine scrubbing is generally employed to capture the CO_2(g) from LNG power plant because it is suitable for large-scale plants (Lim et al., 2021). CO_2(g) is absorbed after gas-liquid contact through an exothermic reaction (Khakharia et al., 2014).

14.2.4 CO_2(g) Liquefaction Process (CLP)

The captured CO_2(g) is liquefied at the CLP and uses a large amount of cold energy to store. The cold energy from the LNG regasification process is recovered to each heat exchanger for CO_2(g) liquefaction (CD3, HX6). The proposed process considered the Linde-dual pressure system for CO_2(g) liquefaction.

14.2.5 Process Model

The developed model of the proposed process for the recovery of waste hot and cold energy in the CCS process in the LNG plant is shown in Fig. 14.1.

The following assumptions were considered to develop the suggested process and Table 14.1 shows the process design based of each feed stream.

14.3 Results and Discussion

14.3.1 The Results of Simulation

To demonstrate energy efficiency of the proposed novel waste energy recovery The LNG power plant for CCS, the total energy consumption of proposed process, and the base process are compared. Table 14.2 shows the results of simulation for base process, and the proposed processes were determined in Table 14.2.

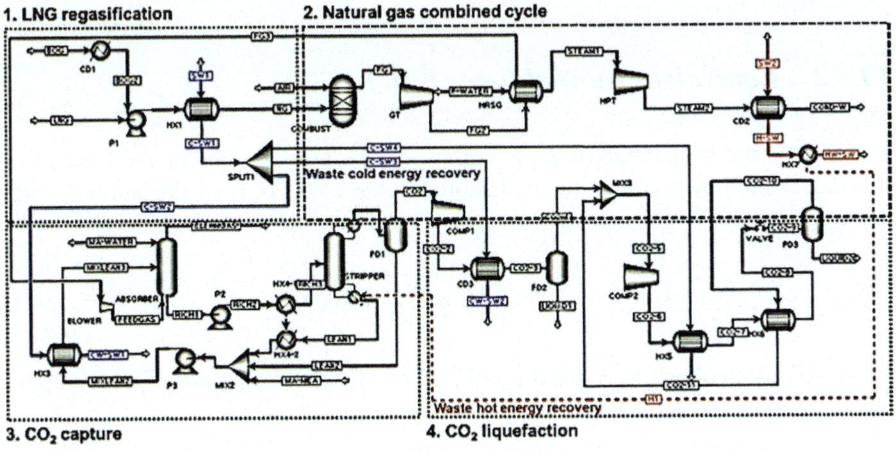

Fig. 14.1 The developed model of the proposed process for recovery of waste energy in CCS process in LNG power plant

Table 14.1 Design basis of each feed stream

Parameter	Values
LNG feed temperature	−162 °C
LNG feed pressure	1 bar
LNG feed mass flow rate	126 tons/h
BOG feed temperature	−50 °C
BOG feed pressure	1.03 bar
BOG feed mass flow rate	0.03 tons/h
Air feed temperature	20 °C
Air feed pressure	1 bar
Air feed flow rate	5.016 bar
Make up MEA* feed temperature	38 °C
Make up MEA* feed pressure	1 bar
Make up MEA* feed flow rate	0.4 tons/h
*MEA = Monoethanolamine.	

Table 14.2 The results of simulation for the base process and proposed process

Parameters		Base process	Proposed process	
LNG regasification	CD1	1.4	1.4	kW
	P1	13,200	13,200	kW
$CO_2(g)$ capture	HX3	13,900	0	kW
	P2	500	500	kW
	P3	3.6	3.6	kW
	S-RB	429,000	241,000	kW
	S-CD	182,000	177,000	kW
$CO_2(g)$ liquefaction	COMP1	28,000	25,000	kW
	COMP2	12,000	11,000	kW
	CD3	38,000	0	kW
	HX3	41,000	0	kW
Total energy consumption		757,605	467,705	kW

14.3.2 *Equivalent Annual Cost (EAC)*

The EAC is the annualized total of capital costs such as equipment and land costs. It is determined by dividing the total capital investment (TCI) by the annuity factor (AF) (Eq. 14.1):

$$\mathrm{EAC} = \frac{\mathrm{TCI}}{\mathrm{AF}} \tag{14.1}$$

where AF is calculated from the discount of the interest rate (r) and number of periods (n) (Lim & Kim, 2021).

TCI is the capital cost, which represents expenses such as land, labor, construction, and equipment costs. It is determined by adding the fixed capital investment (FCI), working capital investment (WCI), and start-up cost (SUC) (Eq. 14.2):

$$TCI = FCI + SUC + WCI \tag{14.2}$$

The base and proposed processes of EAC were determined in Table 14.3.

14.3.3 Total Product Cost (TPC) and Revenue

TPC is the annual cost which is incurred during production. For instance, labor cost, raw material cost, and utility cost. For TPC calculation, herein, the annual operating was set to 365 days. TPC is determined from the direct production costs (DPC) and general expenses (GEs) (Eq. 14.3).

$$TPC = DPC + GEs \tag{14.3}$$

Table 14.3 Total capital investment cost of the base and proposed processes

Parameters	Base process ($/year)	Proposed process ($/year)
Direct cost ISBL		
Equipment cost	140,150,000	138,850,000
Equipment installation	47,350,000	46,110,500
Instrumentation and control	24,950,000	23,246,000
Piping	47,110,000	45,874,000
Electrical	25,237,000	23,962,000
OSBL		
Building and building services	38,975,000	38,012,000
Yard improvements	10,124,000	9,265,000
Services facilities	38,278,000	37,975,000
Land	10,311,000	9,784,000
Total direct cost	372,125,000	371,647,000
Engineering	24,310,000	23,861,000
Construction expenses	24,017,000	23,918,000
Contractor's fee	25,178,000	24,560,000
Contingency	24,569,000	23,235,000
Total indirect cost	93,356,000	92,652,000
FCI	950,758,000	935,873,500
SUC	95,075,800	93,587,350
WCI	190,151,600	187,174,700
TCI	1,235,985,400	1,216,635,550
EAC (r = 5%, n = 15 years)	118,844,750	116,984,187.5

DPC directly affects the production of the product and can be determined from the utility costs (e.g., raw material costs, water costs, and electricity costs) (Eq. 14.4).

$$\mathrm{DPC} = C_{\text{raw materials}} + C_{\text{water}} + C_{\text{electricity}} + C_{\text{maintenance}} + C_{\text{labor}} + C_{\text{supervision}} + C_{\text{operating supplies}} + C_{\text{laboratory}} \quad (14.4)$$

where $C_{\text{raw materials}}$ is raw material costs, C_{water} is water costs, $C_{\text{electricity}}$ is electricity costs, $C_{\text{maintenance}}$ is maintenance costs, C_{labor} is labor costs, $C_{\text{supervision}}$ is supervision costs, $C_{\text{operating supplies}}$ is operating supplies costs, and $C_{\text{laboratory}}$ is the laboratory costs. GEs is the cost incurred as part of the day-to-day operations and can be determined by summing the administrative costs ($C_{\text{administrative}}$), marketing costs ($C_{\text{marketing}}$), and research and development costs ($C_{\text{R\&D}}$) (Eq. 14.5) (Lim et al., 2021).

$$\mathrm{GEs} = C_{\text{administrative}} + C_{\text{marketing}} + C_{\text{R\&D}} \quad (14.5)$$

Revenue is calculated from the produced electricity at gas and steam turbines. The TPC of the base and proposed processes is determined in Table 14.4.

14.3.4 The Results of Economic Evaluation

The comparison of the EAC, TPC, DPC, GEs, revenue, and net profit of the base and proposed processes is determined in Table 14.5.

Table 14.4 TPC and revenue of the proposed and base processes

Parameters	Base process ($/year)	Proposed process ($/year
Direct production costs		
Local taxes, insurance	4,632,000	4,610,000
Maintenance	18,102,000	18,241,000
Operating labor	210,127,000	165,184,000
Supervision and support labor	62,745,000	49,242,000
Operating supplies	2,775,000	2,679,000
Laboratory cost	31,156,000	24,950,000
Plant overhead cost	174,981,000	140,430,000
Water	224,186,000	224,186,000
Electricity	421,148,000	254,012,000
General costs		
Administrative cost	36,215,000	29,654,000
Distribution and marketing cost	154,078,000	122,180,000
R&D cost	49,017,000	37,875,000
TPC	1,389,162,000	1,073,243,000
Revenue	1,775,681,000	1,774,421,000

Table 14.5 The results of economic evaluation

Parameters	Base process ($/year)	Proposed process ($/year)
EAC	118,844,750	116,984,187.5
TPC	524,107,000	467,751,500
DPC	1,149,852,000	883,534,000
GEs	239,310,000	189,709,000
Revenue	1,775,681,000	1,774,421,000
Net profit	314,334,000	630,491,000

14.4 Conclusion

CO_2(g) emissions, global warming, and consequent climate change are serious environmental problems that require scientific attention. In this study, a new process has been studied for the minimization of energy in CO_2(g) capture in a natural gas power plant. Both waste cold energy and hot energy that make up the regasification of LNG; The amine scrubbing system and the CO_2(g) are efficiently recycled to the cooler to lower the temperature of the lean amine solution in the liquefaction process. The fuel usage was reduced by increasing the overall energy efficiency of the LNG power plant for CCS using waste hot and cold energy. With this, it is possible to take a positive environmental protection measure that will reduce global warming by providing economic savings and releasing less CO_2(g) emissions to the atmosphere. The net power generation and exergy efficiency of the proposed process increased by 20% and 11%, respectively, compared to the baseline process. Economic feasibility net profit increased by 78%. It has been observed that the overall energy efficiency and economic feasibility of using waste cold and hot energy have increased, thus reducing fuel use.

References

Bao, J., Lin, Y., Zhang, R., Zhang, N., et al. (2017). Strengthening power generation efficiency utilizing liquefied natural gas cold energy by a novel two-stage condensation Rankine cycle (TCRC) system. *Energy Conversion and Management, 143*, 312–325. https://doi.org/10.1016/j.enconman.2017.04.018

Bao, J., Yuan, T., Zhang, L., Zhang, N., et al. (2019). Comparative study of liquefied natural gas (LNG) cold energy power generation systems in series and parallel. *Energy Conversion and Management, 184*, 107–126. https://doi.org/10.1016/j.enconman.2019.01.040

Khakharia, P., Mertens, J., Vlugt, T. J. H., & Goetheer, E. (2014). Predicting aerosol-based emissions in a post combustion CO_2 capture process using an aspen plus model. *Energy Procedia, 63*, 911–925. https://doi.org/10.1016/j.egypro.2014.11.101

Lee, B., & Lim, H. (2019). Cost-competitive methane steam reforming in a membrane reactor for H_2 production: Technical and economic evaluation with a window of a H_2 selectivity. *International Journal of Energy Research*, 1468–1478. https://doi.org/10.1002/er.4367

Lee, H., Lim, J., Cho, H., & Kim, J. (2022). Novel pulp mill wastewater recovery process for CO_2 and SOx utilization. *Journal of Cleaner Production, 371*, 133298. https://doi.org/10.1016/j.jclepro.2022.133298

Lim J., & Kim J. (2021). Optimizing ash deposit removal system to maximize biomass recycling as renewable energy for CO_2 reduction. *Renew Energy, 190*, 1006–1017. https://doi.org/10.1016/j.renene.2022.03.095

Lim, J., Lee, J., Cho, H., & Kim, J. (2021). Model development of amine regeneration process with electrodialysis reclamation unit. *Computer Aided Chemical Engineering, 50*. https://doi.org/10.3390/min7110207

Lindqvist, K., Jordal, K., Haugen, G., Hoff, K. A., et al. (2014). Integration aspects of reactive absorption for post-combustion CO_2 capture from NGCC (natural gas combined cycle) power plants. *Energy, 78*, 758–767. https://doi.org/10.1016/j.energy.2014.10.070

Nag, P. K., & De, S. (1997). Design and operation of a heat recovery steam generator with minimum irreversibility. *Applied Thermal Engineering, 17*, 385–391. https://doi.org/10.1016/s1359-4311(96)00033-6

O'Neill, B. C., & Oppenheimer, M. (2003). Dangerous climate impacts and the Kyoto protocol. *Science, 302*(80), 1718–1719. https://doi.org/10.1126/science.1071238

Olajire, A. A. (2010). CO_2 capture and separation technologies for end-of-pipe applications—A review. *Energy, 35*, 2610–2628. https://doi.org/10.1016/j.energy.2010.02.030

Paudel, A., & Bandhauer, T. (2018). Techno-economic analysis of waste heat recovery systems for wet-cooled combined cycle power plants. *Applied Thermal Engineering, 143*, 746–758. https://doi.org/10.1016/j.applthermaleng.2018.07.138

Chapter 15
Air Traffic Management Principles: A Case Study on How to Create a Sustainable System

Eva Maleviti

15.1 Introduction

Air traffic management (ATM) relies on integrated, harmonised, and internationally interoperable air transportation. As of international civil aviation organisation (ICAO), the ATM sector needs to go through changes at an operational level with a planning horizon beyond 2025. Air traffic has been significantly growing for the last 20 years; a few years after the COVID-19 pandemic, air transportation started to increase again since people began to fly again. Reliable and safe air transportation must always be maintained, increasing air capacity and using updated air traffic control systems. ATM system must keep up with the fast economic development of aviation worldwide. Passengers need safe, efficient, and affordable air transportation with fewer delays and better airport management. Respect and consideration for the environment are significant to mitigating risks that could compromise aviation safety. These are all necessary elements of a sustainable air transportation system. During flights and air transportation, all users in an inter-operable global air traffic management system must follow specific safety standards, provide and pursue optimum economic operations, meet national security requirements, and aim for sustainable practices.

Air traffic control (ATC) is a system to manage air traffic flow and ensure aircraft safety in the skies. ATC relies on human controllers, technology, and procedures to guide aircraft from take-off to landing. As a complex system, ATC requires a high degree of expertise, skills, and training. However, how the ATM personnel can have a culture towards sustainability? A set of training session have been presented to

E. Maleviti (✉)
College of Aviation, Department of Applied Aerospace Science, Worldwide Campus, Embry-Riddle Aeronautical University, Daytona Beach, FL, USA
e-mail: malevite@erau.edu

T. H. Karakoç et al. (eds.), *Energy and Sustainable Aviation Fuels Solutions*, Sustainable Aviation, https://doi.org/10.1007/978-3-031-70694-3_15

multiple ATC and ATM professionals, as presented below. The current chapter presents these sections in link to sustainability in an effort to bring to the professionals' culture the term sustainability.

15.2 The Link of Sustainability with Air Traffic Management Operations

Air traffic management relies on several principles that define how an efficient ATM system operates. Like many other operations and techniques related to aviation activities, the ATM environment is also driven by strict safety standards and increased commercial and industrial expectations. There are explicit standards for global interoperability, and many States have systems that have evolved and still evolve within frameworks that can sustain their requirements. ATM relies on service provision and requires several resources, airspace, aircraft, humans, and airports. ATM system controls flight between different airports through the airspace, safely and hazard-free, within capacity limits, with optimum use of all system resources. These features and their operations are based on the human factor involvement and capabilities, the whole infrastructure, technologies, interaction, and interrelation of operations. However, as in all aviation operations, ATM needs support to meet the ever-growing global harmonisation and interoperability expectations. Sustainability in ATM will be applied and start from its principles, safety, humans, technology, collaboration continuity, and information. Air traffic management relies on the basic operating principles; safety, humans, technology, information, collaboration, and continuity. Each of these components is vital for the harmonized operations of a sustainable ATM system. Air traffic management requires high safety standards, with distinct processes in a structured and dynamic safety management system. ATM safety management system is necessary for efficient and effective services. Safety in the aviation industry is part of social and economic sustainability. Humans are central to the air traffic management system, as in all aviation operations. To minimise the factors that might compromise safety or economic conditions in the aviation sector, necessary resources, from current training to new technologies, hardware and software and related training, are essential in forming a sustainable working environment. Culture is also a valuable element for any human-centred system. Organisational culture can influence the system towards a sustainable culture, considering what needs to be done from people's verge.

Technology in air traffic management is critical for various operations. Navigation and communication systems, on the ground and air infrastructures, surveillance, data management, and information technology are compulsory to ensure safe, efficient, and effective communication with aircraft while on the ground or air. A system to be sustainable must not only have the necessary people with the necessary skills to work in that environment to preserve all the safety standards but also have all the necessary technology available to preserve their working operations and comply with all the necessary safety standards.

Air traffic management operations depend extensively on timely, relevant, quality-assured information to make informed decisions supporting air transport operations under high safety standards. Information must be shared promptly, efficiently, and accurately, allowing all air traffic management community to conduct their business operations safely and efficiently. Communication is a critical element as far as it concerns a system's sustainability. Aviation is a multicultural and international environment with people from different regions, countries, and with different native languages. Communication and the information transmitted can be crucial elements for making a system operate sustainably; requesting clarification when information is not very well received, trying to use a more apparent accent or speaking slowly, especially when working in stressful conditions or being open to giving clarification or explanation to somebody who is not very fluent in the language used are the factors that can make a system to operate under sustainability aspect. Finally, all aviation systems must apply contingency measures when in high-risk operations. It includes specific processes that can mitigate risks, compromising air traffic safety, such as significant outages, natural disasters, civil unrest or any security threat, and other unusual circumstances. The continuity of operations constitutes a viable and sustainable system (ICAO, 2005).

15.3 Enhanced Performance-Based Navigation

Performance-based navigation (PBN) is a modern approach to air navigation that allows aircraft to fly more precise and efficient routes based on their performance capabilities. Rather than relying on traditional ground-based navigation aids, PBN uses advanced satellite-based technologies such as global positioning systems (GPS) and inertial navigation systems to guide aircraft. The continued growth of air traffic and the need for flight efficiency mandates optimising the available airspace. Exploiting technological advancements in communication, navigation, and surveillance enhances air traffic management operations. Applying area navigation techniques in all flight phases contributes to improved airspace optimisation. Area navigation is enabled using an onboard navigation computer called an area navigation (RNAV) system which maximises the airspace capacity. Flight crew and ATC must apprehend RNAV system capabilities and confirm that these capabilities match the airspace requirements. RNAV systems' use lies at the core of performance-based navigation (PBN). The PBN system allows pilots to fly more direct routes, optimise their flight paths, and reduce fuel consumption, reducing emissions and lower costs. The system also improves safety by reducing the risk of mid-air collisions and allowing aircraft to access previously challenging areas due to terrain or weather conditions.

Aircraft to use PBN have navigation systems that meet specific performance standards and pilots with training and certification to operate the PBN system. Air traffic control must also be equipped with the necessary technology and training to manage PBN routes and procedures. The starting point of performance-based navigation is area navigation (RNAV). The RNAV method permits the aircraft's operation on any desired flight path within the coverage of station-referenced navigation aids, within the limits of the capability of self-contained aids, or a combination of these. With the start of area navigation, there was the requirement to define and standardise the navigation capability, which led to the development of a list of performance-based navigation specifications. The aircraft must remain within a lateral value for each of these specifications to be approved. ICAO Member States, through their National Aviation Authorities, have many reasons and benefits for implementing PBN.

First and foremost, aviation safety is the responsibility of all flights in their territories. Given the increased flight demand, PBN approaches, airspace, and operation approvals will also continue to grow, requiring the involvement and acceptance of the National Aviation Authorities (NAA). The NAAs must fully know PBN to ensure a safe aviation operational environment for all its constituents (ICAO, 2016).

15.4 PBN and Environmental Sustainability

Expanding PBN and aircraft performance, there is a proven connection between them and environmental sustainability. The necessity for improved performance and functionality of navigation systems brought changes to aircraft design, separation minima—reduced vertical separation minima (RVSM), route spacing, airport access, procedure design, and air traffic management. All these changes led to more efficient aircraft operations, reducing greenhouse gas emissions and aircraft noise levels. The PBN concept is an ICAO initiative developed in 2009. The requirement for performance, equipment functionality, and improved infrastructure necessitated the development of new navigation specifications. PBN has a practical implementation in flights, followed by documents, standards, regulations, and certifications. Despite its original creation as a performance navigation tool, it enabled direct and indirect environmental benefits. First, fuel savings is one significant benefit. PBN offers options for shorter flight tracks, optimum profile descent (OPD) capabilities, continuous climb operations (CCO), and continuous descent operations (CDO), also contributing to fuel savings in aircraft flights. In addition, PBN reduces flight variance through more predictable operations. It lowers the landing minima reducing weather cancellations and diversions. One of the benefits of PBN in the fuel savings part is that it also reduces the probability of missed approaches or diversions and airport holdings due to airspace and airport capacity. Finally, PBN reduces the contingency fuel requirements owing to system predictability and reliability. Improving fuel savings leads to reduced CO_2 emissions and other pollutants such as carbon monoxide or other nitrogen oxides (ICAO, 2016).

Additionally, shorter flight paths and vertical profiles allow lower thrust levels with operations with lower noise levels. PBN also offers flight paths to non-noise sensitive areas, with optimum drag profiles that support lower aerodynamic noise. Inevitably, the positive environmental effects are worthwhile. However, the PBN scheme and its operations, expand also to noise reduction, in residential areas or airport facilities and their staff. Consequently, the role of PBN expands to social and environmental sustainability. Considering that ICAO CORSIA and EU-ETS emission schemes are rigorous in emission calculations and that after 2026 the EU carbon taxation will come into force, every scheme and effort to control emissions will be valuable for operators.

15.5 Aircraft and Air Traffic Control Cybersecurity

The increased connectivity of aircraft and air traffic control systems through the federal aviation administration (FAA)'s mandatory NextGen updates to the National Airspace (NAS) open them up to even more dangerous cybersecurity risks. One example is Automatic Dependent Surveillance-Broadcast (ADS-B), which monitors nearly every civil aircraft in the air to ease air traffic control and manage separation between aircraft. Pilots can see the precise locations of any nearby aircraft; air traffic controllers use this data to track aircraft and ensure no potential for spatial conflicts or collisions. One issue might be that since ADS-B operates on unencrypted IP networks and global positioning systems (GPS), It can be vulnerable to outside cyberattacks (Lygkou et al., 2019).

There are three types of attacks on ADS-B and GPS systems. The first type of attack is that GPS signals can be disrupted, making navigation more difficult for pilots and tracking more difficult for controllers. The second type of attack can obstruct communication systems and networks, making air-to-air communication between pilots and air-to-ground communications or between pilots and controllers more difficult or even impossible. The third type of attack is ADS-B communications, which can cause denial of service, or modification or deletion of messages, causing confused communications between operators. Thus, aviation safety is highly compromised. Given the safety implications of missing or inaccurate navigation and locational data, the effect on sustainability must be addressed. Any attack on navigation and communication systems would result in reduced situational awareness by pilots and air traffic controllers (Haas et al., 2016).

The cybersecurity risks for aviation are not confined to airports and airborne aircraft. Commercial operators, cargo carriers, regulators, and aircraft maintenance providers all utilize IP systems for booking, records, and general businesss operations. Without proper cybersecurity measures, these organisations open themselves to attacks on internal financial data, maintenance records, customer data, and more, which can shut down entire networks until they agree to pay a ransom. These scenarios can impact the economic sustainability of airports and air traffic control, also affecting airliners and related or involved stakeholders. Cybersecurity is necessary for the aviation sector since any system that has compromised safety and security operations automatically ceases to be sustainable.

15.6 Conclusion

Finishing the respective training with the above-mentioned topics, ATM professionals, especially ATC staff, showed a positive perception on what sustainability means for their operations and how its principles can be expanded in practice. The holistic approach of aviation sustainability requires all aviation professionals to be informed and engaged on the industry's overall effort. To bring change to a system, it is essential first to understand the basic principles that mandate its functions. Safety, humans, technology, information, and continuity are the basic principles of air traffic operations. Each one of these elements must be maintained and supported so that ATC operations are not compromised. The preservation and proper implementation of technologies show the development of the sector and their contribution to sustainability. PBN is a tool describing an aircraft's capability to navigate using performance standards. However, its effects on environmental sustainability are notable in flight performance, fuel consumption optimisation, and emissions. Finally, the role of cybersecurity in ATM, navigation services, and airport operations is part of cybersecurity and the elimination of cyber risks for sustainability. The trainees found the connection of their everyday operations with sustainability extremely valuable. Often, sustainability is hidden in actions that people already do; however, they must see the connection. Training is always a good start, especially in a newly introduced concept, such as sustainability.

References

Haass, J., Sampigethaya, R., & Capezzuto, V. (2016). Aviation and cybersecurity: Opportunities for applied research. *TR News* (p. 304). https://commons.erau.edu/publication/299

International Civil Aviation Organisation. (2005). *Global air traffic management operational concept*. https://www.icao.int/airnavigation/IMP/Documents/Doc%209854%20-%20Global%20ATM%20Operational%20Concept.pdf

International Civil Aviation Organisation. (2016). *Performance-based navigation (PBN)*. ICAO PBN. Retrieved December 10, 2022, from https://www.icao.int/safety/pbn/miscellaneous%20items/pbn%20faqs.pdf

Lykou, G., Anagnostopoulou, A., & Gritzalis, D. (2019). Smart airport cybersecurity: Threat mitigation and cyber resilience controls. *Sensors, 19*(1). https://doi.org/10.3390/s19010019

Chapter 16
Impact of COVID-19 on Air Cargo Transportation in Turkey

Melike Peynircier and Ali Emre Sarılgan

16.1 Introduction

The significance of air freight continues to grow steadily as a result of the advantages it offers to both customers and industry. The COVID-19 pandemic has further highlighted the indispensability of this entity because of its distinct characteristics (Hava, 2022). Cargo transportation is considered a supplementary revenue stream for aviation enterprises in addition to passenger transportation (Budd & Ison, 2016). However, the primary revenue source has garnered more attention during the COVID-19 pandemic (Bockstaele et al., 2023).

According to Zhang and Zhang (2002), cargo transport operates in a unidirectional manner, implying that air cargo supply is more vulnerable to competitive forces within the industry compared to passenger transport. Furthermore, the restricted availability of cargo airports, which are exclusively designated for the operation of aircraft used solely for cargo purposes, is an additional contributing factor to the variability in the supply of cargo (Kupfer et al., 2011). According to Bombelli et al. (2020), the categorization of airports as either primary (hub) or secondary locations, in relation to the imbalance between supply and demand, influences the quantity of cargo transported.

Cargo flights are designed to transport cargo that has a high volume and low weight, with the objective of optimizing the utilization of the available loading area (Chao & Li, 2015).

According to Lakew (2014), air cargo transportation can be categorized into three distinct classifications: "only cargo," combined," which involves the simultaneous transportation of both passengers and goods and integrated," which includes

M. Peynircier (✉) · A. E. Sarılgan
Institute of Postgraduate Education, Eskisehir Technical University, Eskisehir, Türkiye
e-mail: melikepeynircier@ogr.eskisehir.edu.tr; aesarilgan@eskisehir.edu.tr

T. H. Karakoç et al. (eds.), *Energy and Sustainable Aviation Fuels Solutions*, Sustainable Aviation, https://doi.org/10.1007/978-3-031-70694-3_16

door-to-door delivery services. During the COVID-19 pandemic, enterprises involved in air cargo transportation devised various applications in response to the exceptional circumstances. Air cargo businesses in Turkey implemented several strategies to sustain their operations, mirroring global trends. Consequently, the financial and management circumstances of the firms underwent alterations. The objective of this study is to uncover the operational procedures employed by four air freight businesses that are currently functioning within the borders of Turkey.

16.2 Literature

In their study on the ramifications of the pandemic on the aviation industry, Akca (2020) conducted a comprehensive evaluation of the measures implemented in Turkey, the alterations in operating procedures, the economic implications, and the global consequences. The study conducted by Bakrc (2020) investigated the effects of the COVID-19 pandemic on the air transportation sector in Turkey. In their study, Deveci et al. (2022) conducted an analysis of the effects of the COVID-19 pandemic on the civil aviation sector in Turkey. The researchers investigated various aspects, including the overall market conditions, the implementation of hygiene protocols by industry players, and the financial status of these firms. The study conducted by Nhamo et al. (2020) investigated the effects of the COVID-19 pandemic on the financial performance of airlines.

A sectoral analysis of air cargo transportation during the years 2019–2021 was conducted by Rodoplu and Kanat (2022). According to their statement, while passenger travel experiences negative effects, freight transport is positively impacted. This may be attributed to the urgent demand for medical equipment and the rise of e-commerce. Saban et al. (2021) developed a novel framework aimed at effectively addressing interruptions in cargo capacity and demand imbalances resulting from worldwide disasters, with a particular focus on the impact of COVID-19 on air freight transportation. The study conducted by Maneenop and Kotcharin (2023) aimed to assess and compare the effects of the pandemic and vaccine transportation on the aviation sector.

16.3 Methodology

The current case study is predicated upon the use of semi-structured interviews conducted with a cohort of airline specialists who willingly consented to partake in the research endeavor. The semi-structured interview questions were designed based on expert advice. The second phase of the study was an examination of the airline sector report, the annual report of the State Airports Authority, and the annual report of the General Directorate of Civil Aviation. This was done to validate the information gathered from the interviews.

The absence of publicly available air cargo price data has hindered the ability to conduct a full analysis of the air cargo companies operating in Turkey throughout the pandemic period. Therefore, the study was conducted utilizing data pertaining to cargo and cargo tons transported during the period spanning from 2019 to 2023. Furthermore, the study utilized data from the initial quarter of 2023, while the publication of yearly reports is still pending due to the ongoing nature of the year 2023.

16.4 Findings

The total volume of freight transportation in Turkey experienced a notable increase from 3,855,231 tons in 2018 to 4,090,168 tons in 2019, representing a growth rate of 6.1%. In the year 2020, there was a reduction of 38.89% in domestic flights and 39.97% in international flights, resulting in an overall average decline of 39.11%. Consequently, the total volume of flights decreased to 2,490,521 tons. When comparing the years 2021 and 2019, it is observed that there was a decrease of 16.24% in cargo traffic on domestic routes, a decrease of 16.04% on international routes, and an overall decrease of 16.08% in total. In the year 2022, there was a notable rise in freight traffic across various routes. Specifically, there was a 12.27% increase seen on domestic routes, a 23.59% increase on foreign routes, and an overall growth of 21.29%. Consequently, the total freight traffic reached a substantial volume of 4163.142 tons. In comparison to the time before the COVID-19 pandemic, there was a decline of 5.97% in domestic freight traffic and 3.77% in international freight traffic. However, overall freight traffic experienced an average gain of 1.78%. These findings are based on data from DHMI (2020a, b, 2021a,b, 2022a, b, 2023a, b).

According to the General Directorate of State Airports Authority (DHMI) Airline Sector Report, there was a decrease in air cargo traffic on domestic routes of 22.27%, a loss on international routes of 9.56%, and an overall decrease of 10.10% from the figures recorded in 2019. Upon analyzing the statistics pertaining to the year 2021, it is evident that there was a notable surge in many sectors. Specifically, the domestic line experienced a substantial growth of 61.90%, while the international line observed a comparatively modest gain of 10.17%. In aggregate, the overall growth rate amounted to 12.40% when compared to the pre-COVID-19 period in 2019. In 2022, there was a 1.16% decline in cargo traffic on the domestic line compared to the previous year. However, when compared to the period before the COVID-19 pandemic, there was a significant increase of 60.03%. On the international route, there was a 1.97% decrease in cargo traffic compared to 2021, although it nevertheless experienced a 7.99% gain when compared to the pre-COVID-19 period. According to the data provided by DHMI (2020a, 2021a, 2022a, 2023a), there was a 10.24% rise in cargo traffic in 2022 compared to 2019. However, there was a 1.92% drop in cargo traffic in 2022 compared to 2021. Figure 16.1 displays the comprehensive freight chart for the years 2019–2022.

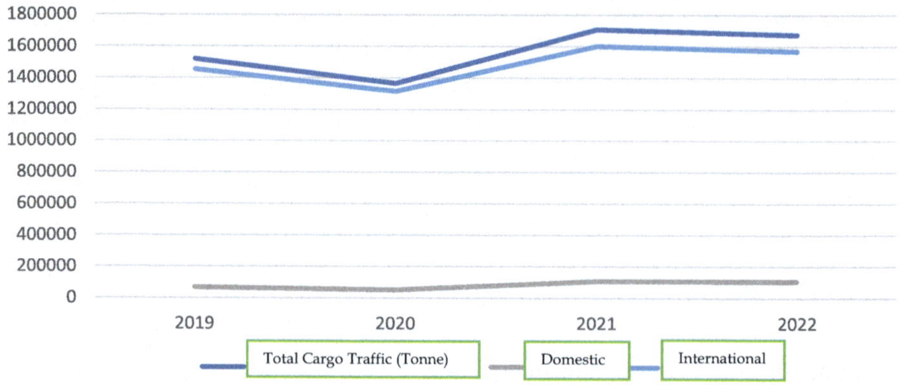

Fig. 16.1 2019–2022 Domestic and international total cargo graph. (**Resource:** DHMI 2023 Airline Sector Report)

As shown previously, overall freight traffic had a decline in 2020, reaching a value of 1,368,577 tons. However, in 2021, it rebounded and reached its peak at 1,711,151 tons. The volume of cargo transportation in the year 2022 exhibits an increase in comparison to the levels observed in 2019, but demonstrates a decline when compared to the preceding year, 2021.

When examining the cargo and freight traffic in Turkey in terms of tonnage from 2019 to 2022, it is hypothesized that the decline in freight traffic observed in 2020 can be attributed to the absence of passenger flights and the subsequent absence of luggage transported with people. In the year 2021, there was a notable surge in cargo traffic due to the growing fleet of aircraft and the continued transportation of medical equipment and vaccines. Additionally, freight traffic experienced a rise in comparison to the previous year as travel restrictions were progressively loosened and passenger flights resumed their regular operations. The observed increase in overall cargo and freight traffic in 2022, surpassing pre-COVID-19 levels, is believed to be attributable to the concurrent expansion of the fleet structure.

In 2019, the Turkish civil aviation industry possessed a fleet of 30 cargo planes. However, the cargo traffic in Turkey accounted for a mere 1.2% of the overall aviation activity during that year. The imposition of travel restrictions at the onset of the year, coupled with the suspension of passenger flights, has also exerted a notable influence. Furthermore, the augmentation in the availability of medical equipment and vaccines during this progression resulted in a surge in the requisition for air transportation of goods. It is hypothesized that the swift recuperation of passenger flights, coinciding with the normalization of cargo plane numbers, led to a 33% rise in cargo flights in 2021, accompanied by a 3% decline in cargo traffic. According to the General Directorate of State Airports Authority (DHMI) (2023a) and Directorate General of Civil Aviation (SHGM) (2023), despite the presence of 35 cargo planes in 2022, it is evident that cargo traffic is approaching levels similar to those observed before the COVID-19 pandemic.

The annual augmentation of cargo aircraft within the fleet has resulted in a corresponding growth in the overall cargo capacity of said aircraft. The capacity of the system had a gradual growth throughout the years. Specifically, it was recorded as 2,296,450 kg in 2019, followed by increments to 2,453,450 kg in 2020, 2,593,450 kg in 2021, and ultimately reaching 2,675,750 kg in 2022 (SHGM, 2020, 2021, 2022, 2023).

16.5 Conclusion

The study evaluated the prices and quantities of cargo transported via phone calls during the period of June 10–17, 2023. It examined how enterprises manage this process based on their respective business models. Additionally, the study investigated whether there are differences between domestic and foreign enterprises in terms of the markets they serve and assessed the rate of change in the size of these enterprises. Based on the information provided by the experts during the interviews, the following findings were drawn:

- The surge in cargo flights can be attributed to the supply of essential medical supplies and vaccines amidst the COVID-19 pandemic. In several instances, the occupancy rate of flights reached a level of 100%, thereby achieving full capacity.
- Contrary to expectations, the aircraft within the firms' fleets have not remained idle. On the contrary, there has been a noticeable growth in fleet size through leasing arrangements, the inclusion of charter flights, and the repurposing of passenger planes for freight transportation.
- Operational challenges were encountered due to the closure of borders in certain nations, prompting enterprises to resort to domestic routes through collaboration with intermediaries.
- The heightened demand experienced during the pandemic resulted in a notable rise in operational revenues.
- During the epidemic, businesses chose to raise the cost of their goods in order to retain customers and enhance their competitive advantage relative to the market average observed prior to the pandemic. However, it is noteworthy that these price hikes were not subsequently passed on to customers until the conclusion of the pandemic.
- During the COVID-19 pandemic, air freight businesses implemented additional steps in their operational processes to safeguard the well-being of their employees. Furthermore, cargo transportation firms augmented their employee count for flights in response to the surge in demand during the pandemic era. Additionally, office staff transitioned to a remote working arrangement.
- The utilization of short-time working allowances was not observed; instead, salaries were augmented in response to enhanced profitability and demanding work conditions. The employees who typically operate in an office setting were transitioned to a hybrid work arrangement that incorporates normalizing measures.

References

Akca, M. (2020). The impact of COVID-19 on the aviation sector. *Eurasian Journal of Research in Social and Economic Sciences*, 45–64.

Bakırcı, M. (2020). The impact of the COVID-19 pandemic on air transportation in Turkey. *Turkish Geographical Review*, 45–58.

Bockstaele, V. V., Dewull, W., & Buyle, S. (2023). Solving the mystery of discrepancies and double counting in air cargo through demand and supply big data analysis. *Journal of the Air Transport Research Society, 1*(1), 81–100.

Bombelli, A., Santos, B. F., & Tavasszy, L. (2020). Analysis of the air cargo transport network from a complex network theory perspective. *Transportation Research Part E, 138*.

Budd, L., & Ison, S. (2016). The role of dedicated freighter aircraft in the provision of global air-freight services. *Journal of Air Transport Management*, 1–7.

Chao, C.-C., & Li, R.-G. (2015). Effects of cargo types and load efficiency on airline cargo revenues. *Journal of Air Transport Management*, 1–8.

Deveci, M., Ciftci, M. E., Akyurt, I. Z., & Gonzalez, S. E. (2022). Impact of the COVID-19 pandemic on the Turkish civil aviation industry. *Sustainable Operations and Computers, 3*, 93–102.

DHMI. (2020a). *2019 Airline sector report*. General Directorate of State Airports Authority, Department of Strategy Development.

DHMI. (2020b). *2019 Annual report*. General Directorate of State Airports Authority, Department of Strategy Development.

DHMI. (2021a). *2020 Airline sector report*. General Directorate of State Airports Authority, Department of Strategy Development.

DHMI. (2021b). *2020 Annual report*. General Directorate of State Airports Authority, Department of Strategy Development.

DHMI. (2022a). *2021 Airline sector report*. General Directorate of State Airports Authority, Department of Strategy Development.

DHMI. (2022b). *2021 Annual report*. General Directorate of State Airports Authority, Department of Strategy Development.

DHMI. (2023a). *2022 Airline sector report*. General Directorate of State Airports Authority, Department of Strategy Development.

DHMI. (2023b). *2022 Annual report*. General Directorate of State Airports Authority, Department of Strategy Development.

Hava, H. T. (2022). Evaluation of the effects of air cargo transportation on global competitiveness. *Journal of Aviation, 6*(2), 206–217.

Kupfer, F., Meersman, H., Onghena, E., & Voorde, E. V. (2011). Air freight and merchandise trade: Towards a disaggregated analysis. *Journal of Air Transport Studies, 2*(2), 28–48.

Lakew, P. A. (2014). Economies of traffic density and scale in the integrated air cargo industry: The cost structures of FedEx express and UPS airlines. *Journal of Air Transport Management, 35*, 100–116.

Maneenop, S., & Kotcharin, S. (2023). Comparative evidence on the impact of the COVID-19 outbreak and vaccine. *Transportation Research: Interdisciplinary Perspectives, 17*, 100741.

Nhamo, G., Dube, K., & Chikodzi, D. (2020). COVID-19 and implications for the aviation sector: A global perspective. In *Counting the cost of COVID-19 on the global tourism industry* (pp. 89–107). Springer.

Rodoplu, H., & Kanat, A. O. (2022). Air cargo industry in the pandemic period: A sectoral research study. *International Journal of Social Sciences*, 2090–2097.

Shaban, I. A., Chan, F., & Chung, S. (2021). A novel model to manage air cargo disruptions caused by global catastrophes such as COVID-19. *Journal of Air Transport Management, 95*, 102086.

SHGM. (2020). *2019 Annual report*. Directorate General of Civil Aviation.

SHGM. (2021). *2020 Annual report*. Directorate General of Civil Aviation.

SHGM. (2022). *2021 Annual report*. Directorate General of Civil Aviation.

SHGM. (2023). *2022 Annual report*. Directorate General of Civil Aviation.

Zhang, A., & Zhang, Y. (2002). A model of air cargo liberalization: Passenger vs. all-cargo carriers. *Transportation Research Part E, 38*, 175–191.

Chapter 17
Progress on PEM Fuel Cell–Powered Unmanned Aerial Vehicle Research

Mustafa Azer, C. Ozgur Colpan, and T. Hikmet Karakoç

Nomenclature

UAV	Unmanned aerial vehicle
PEM	Proton exchange membrane
MTOW	Maximum take-off weight
Li-Po	Lithium polymer

17.1 Introduction

The current state of electric unmanned aerial vehicles (UAVs) shows that flight time is one of the main problems (Çoban & Oktay, 2018). Scientists are trying different ways to solve this problem. Hybrid systems are foremost among these. Solar panels, fuel cells, or batteries with super capacitors are hybridized (Lee et al., 2014). In this study, the development of fuel cell hybrid UAVs is discussed.

The biggest advantage of fuel cell hybrid UAVs is that they can significantly increase flight time by using hydrogen. According to studies in the literature, while the flight time of UAVs with batteries alone is around 1–2 h, this flight endurance can increase to around 20–25 h when fuel cells are included in the system (Swider-Lyons et al., 2011). Proton exchange membrane (PEM) fuel cells play a major role

M. Azer (✉).
Faculty of Aeronautics and Astronautics, Eskisehir Technical University, Eskisehir, Türkiye
e-mail: mazer@eskisehir.edu.tr

C. O. Colpan
Department of Mechanical Engineering, Dokuz Eylul University, Izmir, Türkiye
e-mail: ozgur.colpan@deu.edu.tr

T. H. Karakoç
Faculty of Aeronautics and Astronautics, Eskisehir Technical University, Eskisehir, Türkiye

Information Technology Research and Application Center, Istanbul Ticaret University, Istanbul, Türkiye
e-mail: hkarakoc@eskisehir.edu.tr

T. H. Karakoç et al. (eds.), *Energy and Sustainable Aviation Fuels Solutions*, Sustainable Aviation, https://doi.org/10.1007/978-3-031-70694-3_17

in this difference due to their high energy density (González-Espasandín et al., 2014). However, the power densities of fuel cells are not sufficient to meet the instantaneous high-power requirements of UAVs. Therefore, fuel cell-battery hybrid systems are preferred instead of using a fuel cell as the sole power source. Lithium polymer (Li-Po) batteries can meet instantaneous high-power requirements due to their high-power density (Fig. 17.1).

There are different types of fuel cells such as direct alcohol, solid oxide, alkaline, and proton exchange membrane (PEM). Each type has its own advantages and disadvantages. The prominent feature of PEM fuel cells compared to others is that they can operate at low temperatures such as 20–80 °C and are small in size (Rostami et al., 2022). Although their electrical efficiency is lower than other types, their operating temperature and small size features make them prominent for use in aviation applications.

However, some problems in the current state of fuel cells make them difficult to use in aviation applications. They are costly systems, and another challenge is that it is difficult to achieve stable operating performance in variable atmospheric conditions (Baroutaji et al., 2019).

17.2 Literature Survey on Fuel Cell–Powered UAV Studies

According to the literature review, a large number of studies on fuel cell UAVs have been encountered (Table 17.1). These studies date back to the last years of the twentieth century, and especially after the 2010s, the number of studies has started to increase considerably. Based on the literature, it can be said that this increase is in parallel with the development of the commercial UAV sector and fuel cells.

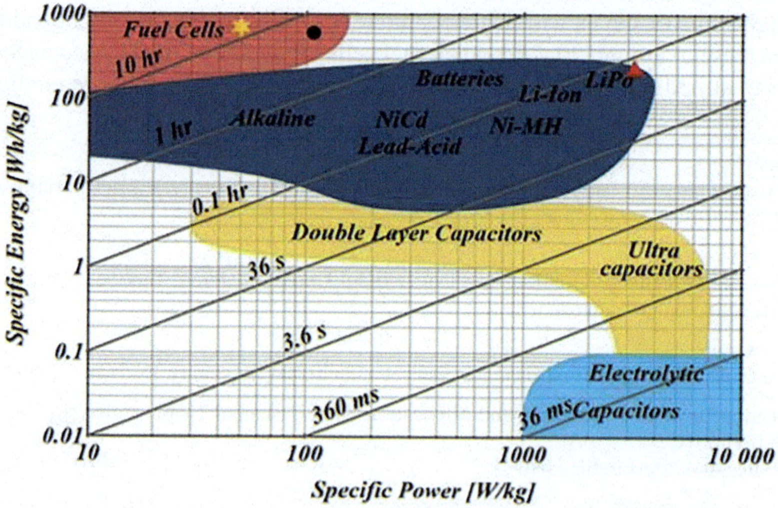

Fig. 17.1 Performances of specific power vs specific energy for different electrical energy-storage technologies. (Boukoberine et al., 2019)

Table 17.1 Fuel cell–powered UAV studies from 2006 to 2023

Year	Demonstrator	Study	UAV type	Power system	FC type	Hydrogen source	FC power	MTOW (kg)	Wingspan	Cruise speed	Test flight
2013	Dudek et al.	Practical	Fixed wing	Lipo + PEM	PEM	High-pressure cylinder	250 W	5.5	3 m	–	30 min
2006	Stroman et al.	Practical	Fixed wing	Only PEM	PEM	Composite cylinder	100 W	2.5	2 m	–	3 h 19 min
2018	Gong et al.	Practical	Fixed wing	Battery + FC + supercapacitor	PEM	Metal hydride tank	150 W	6.5	2.8	17 m/s	35 min
2011	Kim et al.	Practical	Fixed wing	Lipo+PEM serial hybrid	PEM	Sodium borohydride	100 W	2.5	2	–	2 h
2006	Moffit et al.	Practical	Fixed wing	Lipo + PEM	PEM	Metal hydride tank	500 W	19.7	6.7	13.5	No time data
2013	Lee et al.	Practical	Fixed wing	Battery + FC + solar panel	PEM	Chemically generated hydrogen	200 W	18.5	6.9	12	22 h
2014	Kim	Practical	Fixed wing	Lipo + PEM	PEM	Sodium borohydride	500 W	2.2	1.5	8.5	1 h 40 min
2017	Lapana-Rey et al.	Practical	Fixed wing	Lipo + PEM	PEM	Sodium borohydride	200 W	9.6	4.7	17	2 h 17 min
2016	Yang et al.	Practical	Fixed wing	Lipo + PEM parallel	PEM (homemade)	Composite cylinder	877 W	21.2	5	–	40 min
2007	Bradley et al.	Practical	Fixed wing	Only PEM	PEM	Composite cylinder	500 W	16.4	6.6	14.5	No time data
2010	Rhoads et al.	Practical	Fixed wing	Only PEM	PEM	Compressed hydrogen	600 W	13.4	5.5	–	No time data
2012	Swider-Lyons en al.	Practical	Fixed wing	Only PEM	PEM	Compressed hydrogen	550 W	16	5.7	14	26 h
2016	Savaris et al.	Practical	Fixed wing	Lipo + PEM	PEM	Composite cylinder	500 W	13	–	–	6 min

(continued)

Table 17.1 (continued)

Year	Demonstrator	Study	UAV type	Power system	FC type	Hydrogen source	FC power	MTOW (kg)	Wingspan	Cruise speed	Test flight
2020	Ozbek et al.	Practical	Fixed wing	Lipo + PEM	PEM	Composite cylinder	250 W	6.5	2.8	16	–
2020	Depcik et al.	Theoretical	Fixed wing	Li-ion + PEM + HFCE	PEM	Metal hydride tank	250 W	1.9	1.4	13	–
2006	Gong et al.	Theoretical	Fixed wing	Lipo + PEM	PEM	Composite cylinder	200 W	–	–	–	–
2014	Zhou et al.	Theoretical	Fixed wing	Only PEM	PEM	Composite cylinder	1000 W	13.5	3.7	–	–
2022	Suewatanakul et al	Theoretical	Fixed wing	Lipo + PEM	PEM	Compressed hydrogen	650 W	25	4	–	–
2006	Herwert et al.	Theoretical	Fixed wing	Only PEM	PEM	Metal hydride tank	670 W	11.4	5.5	13.4–16.7	–
2018	Hyun oh	Theoretical	Fixed wing	PEM + battery	PEM	Compressed hydrogen	1000 W	7	–	14	–
2008	Chiang et al.	Theoretical	Fixed wing	Lipo + PEM	PEM	Composite cylinder	150 W	5	4	–	–
2012	Verstraete et al.	Theoretical	Fixed wing	Lipo + PEM	PEM	Composite cylinder	200 W	18.2	6	20.5	–
2016	Gadalla et al.	Theoretical	Fixed wing	Lipo + PEM	PEM	Compressed hydrogen	200 W	10.2	3.83	17	–
2022	An et al.	Theoretical	Fixed wing	Lipo + PEM	PEM	Compressed hydrogen	–	25	3.5	25	–
2017	Okumus et al.	Theoretical	Fixed wing	Lipo + PEM	PEM	Boron-based hydrogen	220 W	7.5	2.5	–	–
2022	Coelho et al.	Theoretical	Fixed wing	Lipo + PEM	PEM	Compressed hydrogen	800 W	21.7	4	19.5	–

Here, 12 data belonging to the studies conducted are considered, and these data are listed in a table. The studies conducted through the data are compared, and thus it would be possible to interpret why the researchers made these choices. The given table is a guiding table for all researchers who want to conduct research on this subject in the future, thus paving the way for future studies.

There are a total of 12 different columns in the table. The first of these columns is the year of the study. By looking at the years of the study, it will be possible to observe how systems, designs, and preferences have changed over the years. The second column contains the person(s) who conducted the study. By looking at the number of the group of scientists who conducted the study, it can be understood how much scope the study requires in general, or the similarity of the groups conducting the studies can be examined. The next column contains information on whether the study has been put into practice. When we come to the fourth column, the type of aircraft used or planned to be used in the study is included. These types can be fixed wing, rotary wing, or VTOL platforms. The fifth row contains the basic elements of the system planned to power the aircraft. Here, the systems can be single, hybrid, or triple hybrid. In column 6, the type of fuel cell used in the study is listed. Considering the position of fuel cells today, it can be said that there are more than ten types. In the seventh row, information is given about the method of supplying the fuel to be used by the fuel cell. We can say that there are basically two different methods here; one of them is to produce hydrogen in advance and store it in a hydrogen tank, and the other is to include the hydrogen generator in the power system to produce oxygen instantaneously in the required amount at the time of the event. Examples of both methods can be seen in different studies. If the hydrogen storage method is used, it is necessary to choose a fuel tank type. Options include metal hydride hydrogen tanks or composite hydrogen tanks. The next line shows the nominal power rating of the fuel cell. The choice of fuel cell is based on the power requirement of the total system installed. In other words, it would not be wrong to say that the more powerful the fuel cell, the more power the system needs. However, there is one element that can mislead us here: how many power sources the system consists of. If the system is hybrid or triple hybrid, the amount of power required per system will decrease as the energy needed will be divided into these systems. In the ninth column is maximum take-off weight (MTOW). The maximum take-off weight shows us the maximum weight that an aircraft can take off from the ground due to its power system and design. By looking at the weight of the aircraft in the study, it can be understood how big or how light a structure is used. In addition, the link between MTOW and other columns can be examined and the differences in level and purpose between the research groups can be interpreted by making comparisons such as what size and weight of aircraft other researchers are able to manufacture and how low power the aircraft with what weight can be operated with. The tenth column contains information on the wingspan of the aircraft. Wingspan refers to the distance between the wing tips when the aircraft is viewed from above. As the wingspan increases, the lift force obtained from the wings will increase in the right proportion, so it can be interpreted how high lift force is needed in the aircraft. In addition, the required lift force increases in direct proportion to the speed and density of the air flowing over the wing of the aircraft. Here, it can be interpreted how fast the aircraft should fly and in which altitude range it should fly. The penultimate column, the cruise speed column, contains information on the cruise speed that the designed aircraft will use according to its flight profile. Here, since the UAV's ability to fly slowly will increase its functionality, it is desirable that the cruise speed can be low when necessary. However, requirements may vary from mission to mission. When we come to the last column, it is stated whether a test flight was made with the aircraft with the specifications given, and if so, how long it was in the air at one time. We can say that the presence of a test flight in this column and the duration of the flight, if any, will increase in direct proportion to the success of the study.

17.3 Conclusions

In this study, fuel cell UAV studies starting from 2006 until the end of 2022 have been examined in detail, and a large table has been prepared according to the system features. The table is considered to be a good guide for researchers who will work on fuel cell–powered UAV subjects. Based on the conclusions drawn, the concept of a fuel cell UAV is a realistic concept, but some technological and scientific advances need to be made before it can be commercialized. Considering the increasing interest of scientists in fuel cell UAVs in recent years, it would not be wrong to think that these problems will be solved soon.

The study is good preliminary research for the fuel cell Hydra aircraft project planned to be carried out in the future. The results of the project will be shared soon.

References

Baroutaji, A., Wilberforce, T., Ramadan, M., & Olabi, A. G. (2019). Comprehensive investigation on hydrogen and fuel cell technology in the aviation and aerospace sectors. *Renewable and Sustainable Energy Reviews, 106*, 31–40.

Boukoberine, M. N., Zhou, Z., & Benbouzid, M. (2019). A critical review on unmanned aerial vehicles power supply and energy management: Solutions, strategies, and prospects. *Applied Energy, 255*, 113823.

Çoban, S., & Oktay, T. (2018). Unmanned aerial vehicles (UAVs) according to engine type. *Journal of Aviation, 2*(2), 177–184.

González-Espasandín, Ó., Leo, T. J., & Navarro-Arévalo, E. (2014). Fuel cells: A real option for unmanned aerial vehicles propulsion. *The Scientific World Journal*, 2014.

Lee, B., Park, P., Kim, K., & Kwon, S. (2014). The flight test and power simulations of an UAV powered by solar cells, a fuel cell and batteries. *Journal of Mechanical Science and Technology, 28*(1), 399–405.

Rostami, M., Dehghan Manshadi, M., & Afshari, E. (2022). Performance evaluation of two proton exchange membrane and alkaline fuel cells for use in UAVs by investigating the effect of operating altitude. *International Journal of Energy Research, 46*(2), 1481–1496.

Swider-Lyons, K., Stroman, R., Page, G., Schuette, M., Mackrell, J., & Rodgers, J. (2011). Hydrogen fule cell propulsion for long endurance small UVAs. In *AIAA centennial of naval aviation forum 100 years of achievement and progress* (p. 6975).

Chapter 18
The Importance of Exergy for Sustainability in Aviation

Cengizhan Gök, Umut Çiçek, Selçuk Ekici, Özgür Ballı, and T. Hikmet Karakoç

18.1 Introduction

Thermodynamic analysis can be used to make calculations that will improve the efficiency of a system, engine, machine, as well as a method with a wide area used to analyze a nutrient, a chemical mineral or human body and organs. In addition, thermodynamic analysis can be used to better understand any process, facilitate design and control, identify necessary points for process improvement, provide final improvement, determine the degree of perfection of energy use of certain components, parts or parts, and calculate their efficiency.

A portion of the total energy enters a framework, friction, combustion reaction, entropy generation, etc. in the system. It disappears due to irreversibility, and the

C. Gök (✉)
Jet Overhaul Department, 1st Air Maintenance Factory Directorate (1.HBFM), General Directorate of Military Factories (AFGM), Ministry of National Defense (MND), Eskisehir, Türkiye

U. Çiçek
Department of Human Resources, 1st Air Maintenance Factory Directorate (1.HBFM), General Directorate of Military Factories (AFGM), Ministry of National Defense (MND), Eskisehir, Türkiye

S. Ekici
Department of Aviation, Iğdır University, Iğdır, Türkiye

Ö. Ballı
Quality Assurance of Fixed Wing Air Vehicle, Turkish Aerospace Industries Inc. (TAI), Ankara, Türkiye

T. H. Karakoç
Department of Aircraft Airframe Powerplant Maintenance, Eskisehir Technical University, Eskisehir, Türkiye

Information Technology Research and Application Center, Istanbul Ticaret University, Eskisehir, Türkiye

T. H. Karakoç et al. (eds.), *Energy and Sustainable Aviation Fuels Solutions*, Sustainable Aviation, https://doi.org/10.1007/978-3-031-70694-3_18

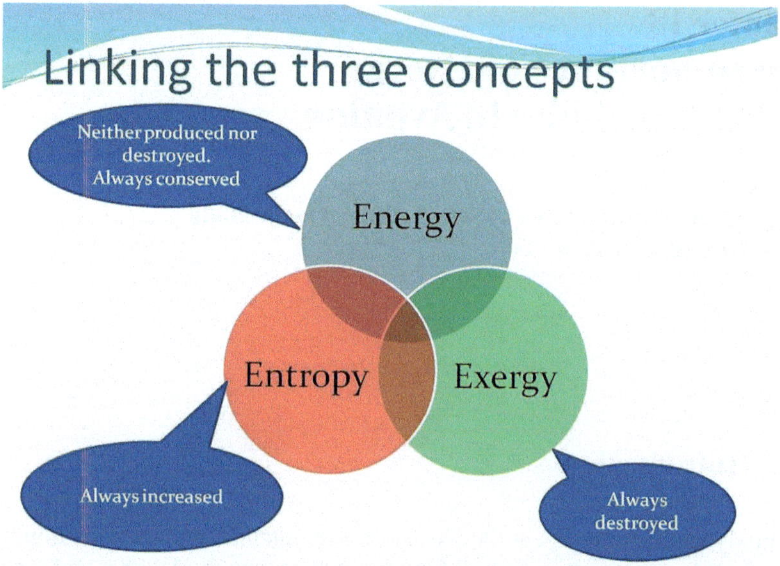

Fig. 18.1 3E: Energy, exergy, entropy

quality of the input energy, that is, its ability to do work, decreases. Therefore, in many bibliographies, the second law is called usable energy, usability, useful energy or exergy analysis (Çengel & Boles, 1996).

The first law of thermodynamics advocates the conservation of energy. The energy analysis gives information about the system in general; it does not take into account the losses caused by the irreversibility (entropy generation) in the framework. The second law of thermodynamics overcomes the fall of energy, the generation of entropy, and the ability to do work during a process. Except in reversible systems, exergy is not conserved like energy. A part of the exergy is destroyed due to irreversibility in the system, and a part of it is thrown out from the system boundaries to the environment exergy loss (Aydın, 2012).

The first and second laws are used to evaluate the productivity of gas turbine aircraft engines used in aircraft. While the efficiency of the engine is examined with the first law analysis, the improvement rate of the engine and the maximum capacity it can reach are determined with the second law (Ekici et al., 2017) (Fig. 18.1).

18.2 Literature Review

Ballı (2023) evaluated the exergy-based peripheral, sustainability, peripheral impact and peripheral loss cost formations of a turboshaft engine used in a helicopter. Although the energy efficiency was found to be 25.18% in his study, the exergy efficiency was obtained as 23.72%.

Aydın (2012) performed an exergetic analysis of a turbofan engine. To separate this acquisition from previous studies; The world's most common engine, the CFM56-7B, was chosen. This engine has not been studied before. Thus, the exergy and sustainability analysis on this engine is presented as a novice study. It is also a first on exergy-based sustainability indicators for airplane engines. Six exergy-based sustainability factors of turbofan propulsion, which are high bypass aircraft engine family members with a thrust range of 86.7–121 kN, have been calculated effectively. In this situation, exergy expenditure, waste exergy ratio, exergy destruction factor, correction effect factor and exergetic sustainability index. This framework was established as the most irreversible unit combustion chamber.

Ekici et al. (2017) discussed performance evaluation parameters for turbojet engines. In recent years, aviation industry and airline carrying have been developing and progressing rapidly gradually. This situation leads to an increase in both the energy consumption and environmental effects of aircraft. Therefore, the productivity of turbojet engines used in many aircraft, especially airplanes, is an issue that needs to be examined. On the other hand, thrust efficiency, thermal efficiency, total efficiency and exergy efficiency, which are other performance parameters of the engine, were determined as 57%, 45%, 25% and 8.05%, respectively. The performance evaluation of gas turbine engines, which are widely used in aircraft, is of great importance in operating processes. The performance parameters to be obtained as a result of processing the data received from the engines give an idea about the maintenance and revision requirements of the engine.

18.3 Engine Exergy Analysis

In the exergy analysis, calculations were made by considering the air in the inlet and compressor part as fluid, combustion chamber, gas turbine, power turbine and exhaust gases. The system elements to be examined in these analyses are as follows:

- Compressor
- Combustion chamber
- Gas turbine
- Exhaust

Exergy assay uses the exergy worth at the inputs and outputs of the system elements. The exergy flows of these stations will be calculated by using the temperature, pressure and flow rate values at the entrances and exits of the stations to be analyzed.

Thermodynamic analysis of the engine was made according to the following assumptions (Fig. 18.2):

- Chemical exergy, except for the combustion chamber, is not included in the calculations.
- It is regarded that there is complete combustion in the combustion chamber.

Fig. 18.2 Airplane picture

- Kinetic energy and exergy, potential energy and exergy changes are negligible.
- The framework is always on with continuous flow.
- It is assumed that complete combustion has taken place.
- System elements are adiabatic.

18.4 Conclusion

Exergy-based sustainability for airplanes is proposed to calculate and label key indicators. This approach will enable both aircraft manufacturers and airlines to assess their sustainability in a standardized way, allowing airline companies to make informed decisions when purchasing and operating existing aircraft, based on these indicators. Exergy-based sustainability indicators will be included as essential criteria. Incorporating exergy into the evaluation of sustainability indicators for airplanes and engines is considered highly beneficial, especially for airline operators and aviation authorities. In general, the efficiency of combustion chambers tends to be low compared to other engine components. When designing next-generation engines, the use of advanced materials, coatings, and configurations aimed at improving combustion and enhancing combustion chamber efficiency will be prioritized. Therefore, research in these areas is expected to yield valuable improvements.

References

Aydın, H. (2012). *Development of exergetic sustainability indicators on commercial PhD thesis Civil Aviation Major Science*.

Balli, O. (2023). *Exergetic, sustainability and environmental assessments of a turboshaft engine used on helicopter*. https://doi.org/10.1016/j.energy.2023.127593

Çengel, Y. A., & Boles, M. A. (1996). *Mühendislik yaklaşımıyla termodinamik*. Literatür Yayıncılık.

Ekici, S., Şöhret, Y., & Karakoç, T. H. (2017). *Performance evaluation parameters for turbojet engines*. https://doi.org/10.30518/jav.323866

https://dergipark.org.tr/tr/download/issue-full-file/44808

https://www.linkedin.com/pulse/so-what-exergy-how-does-affect-our-every-day-life-john-papageorgiou-/

Chapter 19
Operation of New-Generation Aircraft in the Emergency Response Service

D. Szilágyi and D. Sziroczák

Nomenclature

EMS	Emergency medical service
eVTOL	Electric vertical take-off and landing aircraft
UAM	Urban air mobility
API	Application programming interface
EASA	European Union Aviation Safety Agency

19.1 Introduction

The timely response in case of an emergency is an important issue in every country. Locations further away from major cities are often troublesome to reach in the prescribed 15-min time frame. While technological and medical advances made it possible to provide a higher level of healthcare, the increased traffic and workload of healthcare workers decreased the average response time in many areas. The aim of this chapter is to investigate the feasibility of concept that integrates eVTOL (electric Vertical Take-off and Landing) aircraft into the emergency response service in Hungary. Work on the topic of eVTOLs in EMS operations suggests that they have great potential (Chappelle et al., 2018; Mihara et al., 2021; ADAC Luftrettung-Volocopter, 2020). While helicopter air ambulance is available, the associated costs restrict deployment to the most urgent cases. eVTOLs have the potential to make air emergency response more accessible while keeping sustainability in mind.

The proposed concept with the distribution of eVTOL stations across the country is depicted in Fig. 19.1. There are 30 eVTOL stations that are selected from a pool of existing 255 stations optimally to ensure maximum coverage, denoted by the blue

D. Szilágyi (✉) · D. Sziroczák
Faculty of Transportation Engineering and Vehicle Engineering, Department of Aeronautics and Naval Architecture, Budapest University of Technology and Economics, Budapest, Hungary
e-mail: dszilagyi@edu.bme.hu; sziroczak.david@kjk.bme.hu

T. H. Karakoç et al. (eds.), *Energy and Sustainable Aviation Fuels Solutions*, Sustainable Aviation, https://doi.org/10.1007/978-3-031-70694-3_19

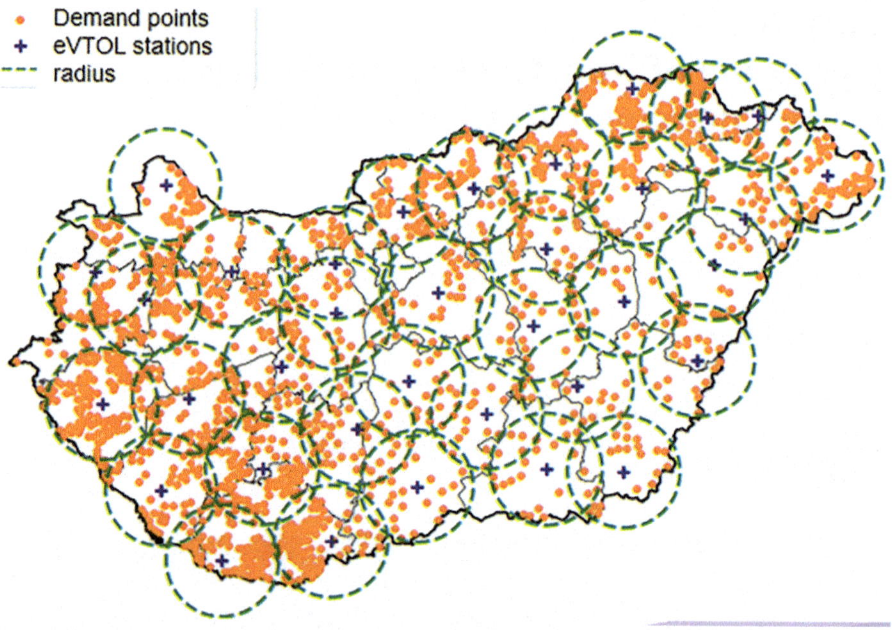

Fig. 19.1 The location of the proposed eVTOL stations in Hungary

crosses in the figure. The operational radius is 34 km, denoted by the green dashed circles, and the demand points, denoted by orange dots, are settlements that need to be covered.

The current calculated response times of the country considering only ground ambulances are shown in Fig. 19.2. The response time for a settlement is represented by the approximated drive time to the centre of the settlement from the nearest ambulance station using the open source OpenStreetmap based routing application programming interface (API) (OSRM, 2022).

If the response times to a settlement are weighted by their population, then the response time is under 15 min on average around 82% of the time. Unweighted, all settlement response time on average is under 15 min only about 66% of the time. This shows the disadvantage of smaller and more remote locations compared to more populous areas. According to the estimate of the proposed concept shown in Fig. 19.2, this potentially could be raised to 99% with eVTOLs, in the ideal case of good weather and exact location information.

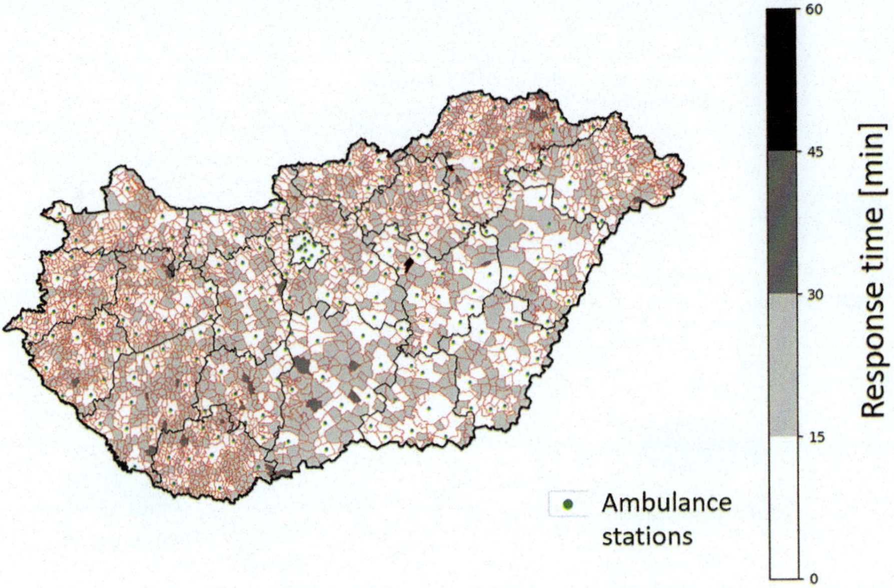

Fig. 19.2 The calculated current ambulance response times in Hungary

19.2 Analysis

The aim of the analysis was to approximate the required performance of the vehicles that would be capable of performing these missions. Based on the draft of upcoming European Union Aviation Safety Agency (EASA) regulations (EASA NPA, 2022), the minimum energy requirements are comprised of the following:

- Trip energy.
- Energy to an alternate destination.
- Contingency fuel.
- Final reserve fuel.

The mission used to determine the required equivalent range of the mission is illustrated in Fig. 19.3. The trip energy is the energy required for a 34 km round trip, the alternate energy was assumed to be one operating radius to guarantee an alternate landing spot, and both the go-around reconnaissance manoeuvre and the final reserve were assumed to be 5 min of cruising equivalent energy. The required speed to make the one-way trip in 15 min was calculated to be 204 km/h, assuming 1–1 min for take-off and landing and on average 3 min for reconnaissance and relocation for nearby landing spot, leaving 10 min for the 34 km radius.

The total required range works out to be 131 km, with three times the radius, 10 min reserve and adding a 5% contingency extra energy for the trip. Based on an interview with the air ambulance service, the required payload capacity was

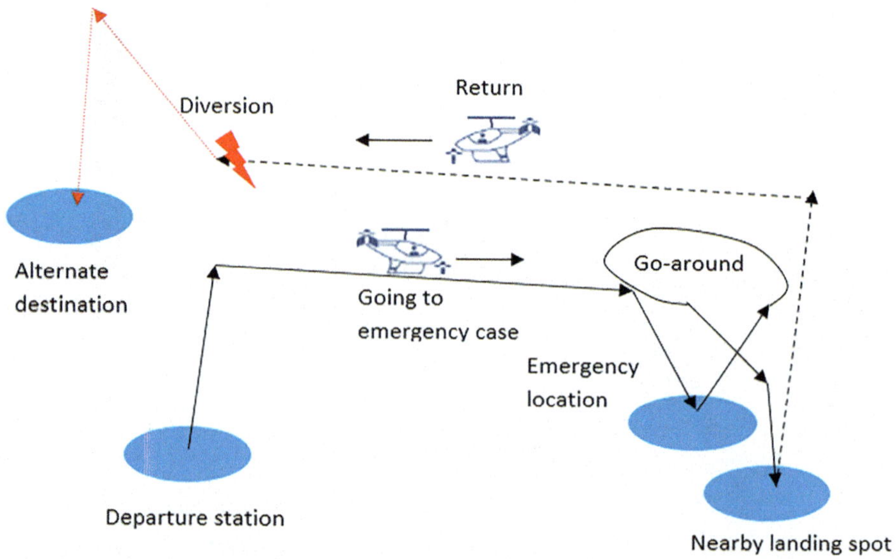

Fig. 19.3 The defined mission of the emergency response eVTOL

determined to be around 410 kg, consisting of the weight of the equipment and three passengers:

- The pilot.
- The doctor.
- Paramedic/technical crew.

This concept considers medical personnel transportation to the scene, but patient transportation was not investigated. In order to determine how close are the requirements to being feasible, the range of 5 eVTOL concepts was estimated using the performance data and method described by Sripad and Viswanathan (2021) and Yang et al. (2021).

These eVTOL concepts are:

- Joby S4.
- KH heaviside.
- Lilium jet.
- Beta alia.
- Archer maker.

The cruising altitude was chosen to be 350 m, and the battery weight of the vehicles was estimated assuming a 40% battery weight fraction. Some concepts are estimated to have a higher, around 50%, battery weight fraction (Bacchini & Cestino, 2019), but it was kept at a more moderate 40% for all vehicles due to the lack of reliable battery weight information. The optimal fraction, if technology allows it, would be around 2/3 or 66% (Traub, 2016). As today's best technology, 240 Wh/kg of battery energy density was considered based on Jaephil et al. (2015). The weight allocated to one passenger was assumed to be 100 kg.

19.3 Results

The ranges of the concepts are depicted in Fig. 19.4 as a function of battery energy density. The possible ranges depending on the payload weight are inside the polygons corresponding to the colour of the aircraft. The lower boundary pertains to pilot-only operation, while the upper boundary describes full occupancy.

The limiting values of range and carried energy assuming 240 Wh/kg maximum energy density are shown in Table 19.1 with full and pilot-only occupancy. The lilium jet performs really poorly compared to the others; this is mainly due to its sensitivity to the length of the vertical climb. Also, the efficiency of the ducted fans of the lilium jet might be underestimated by the used formulae.

Three of the investigated five concepts, Beta alia, Joby S4 and KH heaviside achieved the necessary 131 km range required, with the first two having enough payload capacity as well. However, the strong dependence of feasibility hinging on a quick transition to the more efficient fixed-wing horizontal flight mode is true for all of the concepts, and the energy consumption is much higher during vertical flight. The results indicate that the feasibility of eVTOL application in the emergency service might just be within reach using the best technology available today.

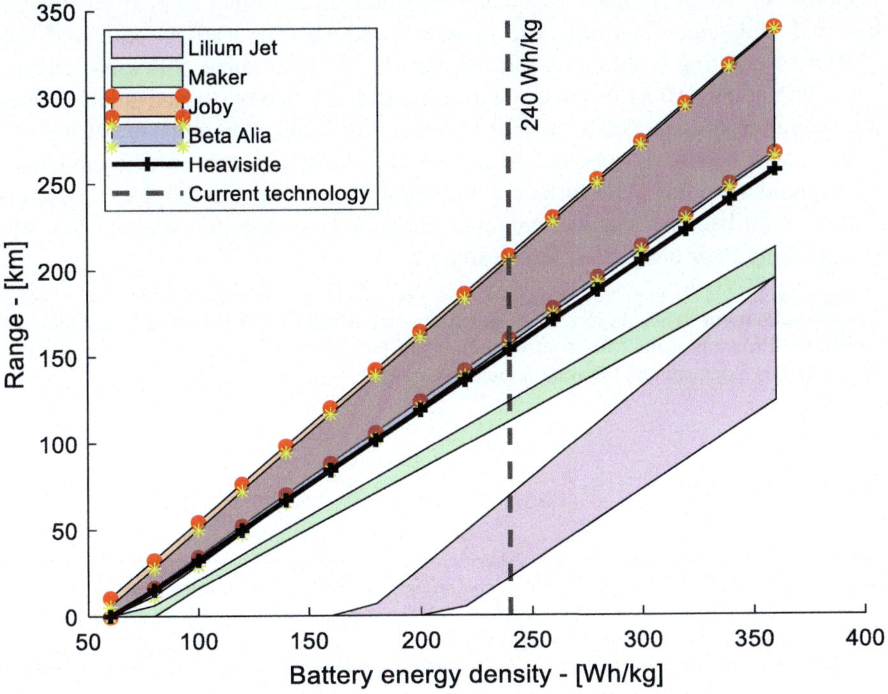

Fig. 19.4 The ranges of the eVTOL concepts as a function of battery energy density

Table 19.1 The carried energy and the range of the concepts at 240 Wh/kg

	Lilium full	Lilium 1 pax	Beta alia full	Beta alia 1 pax	Joby full	Joby 1 pax	Maker full	Maker 1 pax	Heaviside
Total energy - kWh	237.6	237.6	204.2	204.2	161.5	161.5	125.6	125.6	28.3
Passengers	7	1	6	1	5	1	2	1	1
Range - km	22.1	69.5	156.6	205.4	159.9	208.1	112.2	123.9	153.4

19.4 Conclusion

All in all, eVTOLs have the potential to significantly improve the emergency response service in Hungary, especially in remote areas further away from city centres and ambulance stations. The operation required 30 stations equipped with eVTOLs operating within a 34 km radius, 131 km total range, 204 km/h cruising speed and about 410 kg of payload capacity, and with today's best available batteries, the range is estimated to be feasible. Of course, the economic side of the integration also needs to work out in favour of eVTOLs to achieve the envisioned widespread adoption. With further technological improvements, slightly longer distances might become feasible, reducing the number of required stations and vehicles, making their integration less costly.

Acknowledgement This project is supported by the **ÚNKP-22-3-I-BME-57** New National Excellence Program of the Ministry for Culture and Innovation from the source of the National Research, Development and Innovation Fund.

References

ADAC Luftrettung-Volocopter. (2020). *Multicopter in the rescue service: Feasibility study on the application potential of multicopter as emergency doctor shuttles*. Result report.

Bacchini, A., & Cestino, E. (2019). Electric VTOL configurations comparison. *Aerospace, 6*, 26. https://doi.org/10.3390/aerospace6030026

Chappelle, C., et al. (2018). *Opportunities to enhance air emergency medical service scale through new vehicles and operations*. In Aviation technology, integration, and operations conference, 2018.

EASA NPA. (2022). *European union aviation safety agency notice of proposed amendment 2022–06 in accordance with Article 6 of MB Decision No 1–2022 TE.RPRO.00034–011*. https://www.easa.europa.eu/en/downloads/136705/en

Jaephil, C., et al. (2015). Commercial and research battery technologies for electrical energy storage applications. *Progress in Energy and Combustion Science, 48*, 84–101. ISSN 0360–1285. https://doi.org/10.1016/j.pecs.2015.01.002.

Mihara, Y., & Pawnlada, P. et al. (2021). *Cost analysis of eVTOL configuration design for an air ambulances system in Japan*. In CESUN conference. https://doi.org/10.5703/1288284317285.

OSRM. (2022). *Open-source routing machine API*. https://project-osrm.org/

Sripad, S., & Viswanathan, V. (2021). The promise of energy-efficient battery-powered urban aircraft. *Proceedings of the National Academy of Sciences, 118*(45), e2111164118.

Traub, L. (2016). Optimal battery weight fraction for maximum aircraft range and endurance. *Journal of Aircraft, 53*, 1–3. https://doi.org/10.2514/1.C033416

Yang, X., et al. (2021). Challenges and key requirements of batteries for electric vertical takeoff and landing aircraft. *Joule, 5*. https://doi.org/10.1016/j.joule.2021.05.001

Chapter 20
The Impact of SAF on Reducing NO_x, SO_2, and Non-CO_2 Emissions

Mhd Bashar Al Kazzaz and Beneda Károly

Nomenclature

CAF	Conventional aviation fuel
CO	Carbon monoxide
CO_2	Carbon dioxide
EASA	European Union Aviation Safety Agency
EU	European Union
HC	Hydrocarbons
IATA	International Air Transport Association
ICAO	International Civil Aviation Organization
NO_x	Nitrogen oxides
PM	Particulate matter
SAF	Sustainable aviation fuel
SO_2	Sulphur oxides

20.1 Introduction

Carbon dioxide (CO_2), nitrogen oxides (NO_x), sulphur oxides (SO_2), unburnt hydrocarbons (HC), carbon monoxide (CO), particulate matter (PM), and soot are the primary pollutants emitted by the turbojet engines, as shown in Fig. 20.1.

In addition to the primary impact of using SAF, which reduces CO_2 emissions using the SAF production cycle, SAF has a significant climate impact by reducing NO_x, SO_2, PM, and soot.

M. B. Al Kazzaz (✉) · B. Károly
Faculty of Transportation Engineering and Vehicle Engineering, Department of Aeronautics and Naval Architecture, Budapest University of Technology and Economics, Budapest, Hungary
e-mail: balkazzaz@edu.bme.hu; beneda.karoly@kjk.bme.hu

T. H. Karakoç et al. (eds.), *Energy and Sustainable Aviation Fuels Solutions*, Sustainable Aviation, https://doi.org/10.1007/978-3-031-70694-3_20

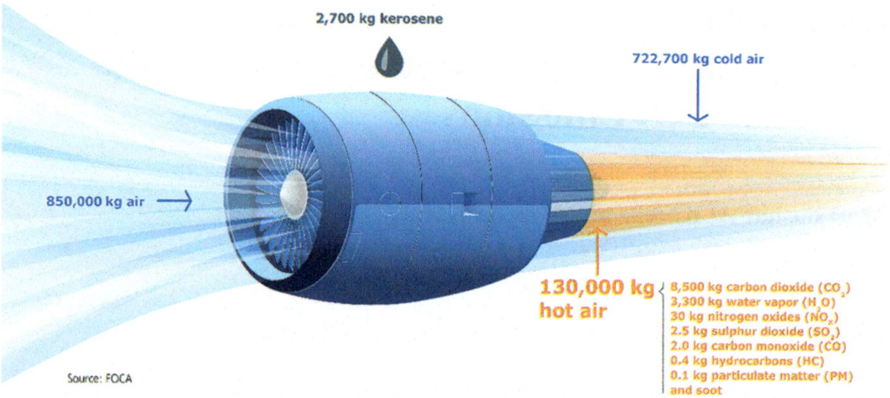

Fig. 20.1 Emissions from a typical two-engine jet aircraft during a 1-hour flight with 150 passengers. (EASA, 2019)

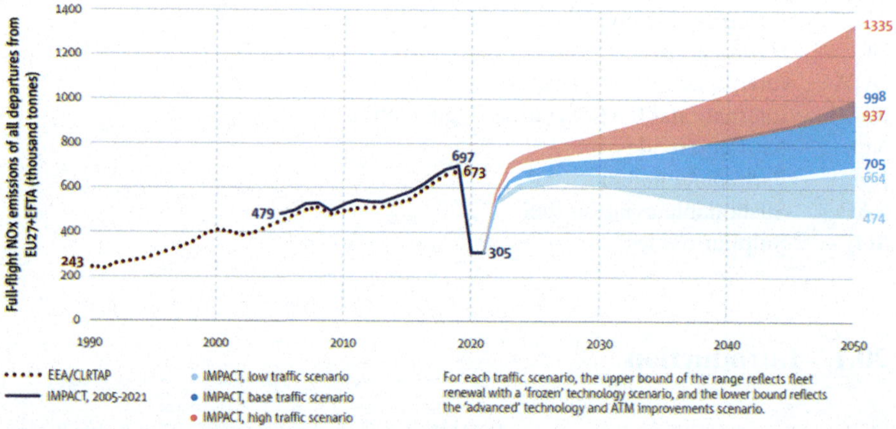

Fig. 20.2 NO_x emissions reached about 700 thousand tonnes

Because SAF has a somewhat different underlying composition than conventional aviation fuel (CAF), its use could impact the balance of these non-CO_2 pollutions.

20.2 NOx Emissions

As indicated in Fig. 20.2, NO_x emissions have increased much more rapidly since 2005, respectively, by 46% and 40% in 2019 and 2017.

Also, we can notice that in the case of the base scenario, the NO_x emission could be 998 thousand tonnes by 2050 (EASA, 2023).

The combustion of SAF results in lower greenhouse gas (GHG) emissions than CAF due to a lower carbon content of the fuel. Figure 20.3 shows the NO_x emission

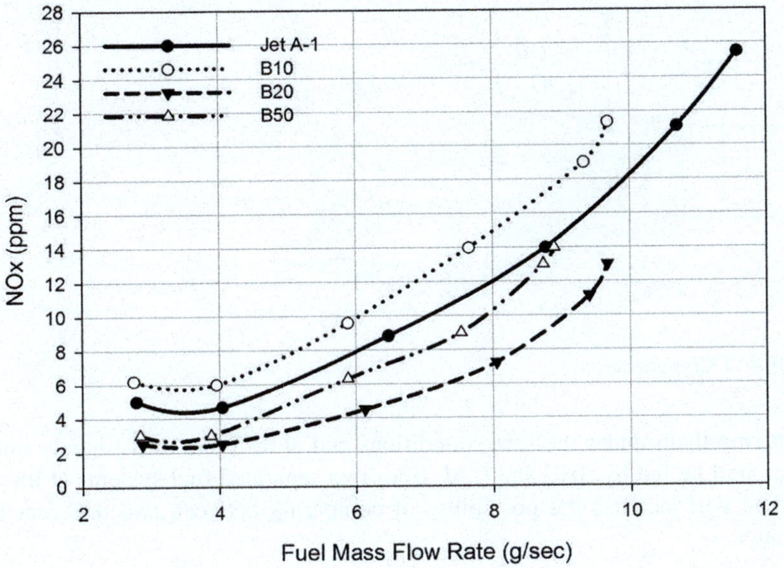

Fig. 20.3 NOx emissions for waste cooking oil biodiesel blends and Jet A-1 at different fuel mass flow rates. (Noureldin et al., 2015)

for different waste cooking oil biodiesel blends and Jet A-1 and their different mixes at various throttle valve settings. The reduction in NO_x emissions may be attributable to the lower heating value of waste cooking oil biodiesel fuels, which results in a lower flame temperature. Overall, NO_x emissions were lowered by increasing the amount of waste cooking oil biodiesel in the blend (Noureldin et al., 2015).

In the previous research, the blending ratios of the waste cooking oil biodiesel and the Jet A-1 were blended at 10% (B10), blend 20% (B20), and blend 50% (B50) on a volume basis.

20.3 Analysis of SAF Emissions Experimentally on TKT-1

This TKT-1 engine is a single spool engine (Fig. 20.4) which is located on a test bench at the Department of Aeronautics and Naval Architecture of Budapest University of Technology and Economics; it has a centrifugal compressor, axial turbine, and annular combustion chamber which has four nozzles that inject the required fuel. Currently, these four nozzles are sourced by one fuel system. A modification to the fuel system will be done by installing a new additional fuel system to the engine to feed one nozzle with SAF and the remaining three nozzles will stay fed with CAF.

After doing this modification on the TKT-1, we will be able to analyse and investigate the differences between SAF and CAF emissions experimentally by

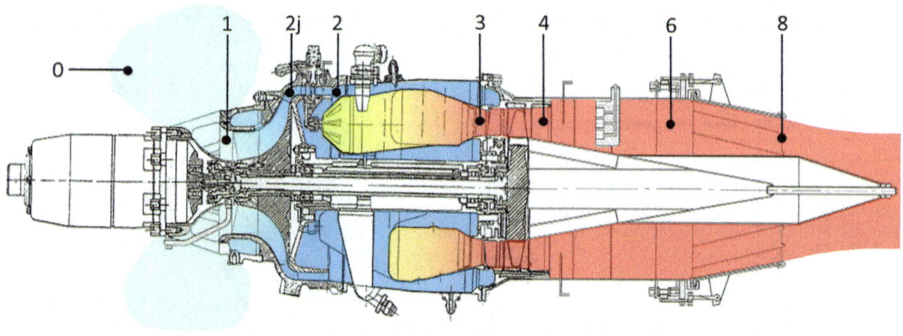

Fig. 20.4 TKT-1 engine

comparing them under the same conditions and at different thrust levels since the engine will be fed by SAF and CAF from two separated fuel systems at the same; all these will increase the possibility of comparing between two different blending ratios.

20.4 Conclusion

SAF has proven to be a promising solution for reducing aviation non-CO_2 impacts, which are mostly caused by NO_x, SO_2, and soot emissions, and has demonstrated its potential to mitigate environmental impacts and improve air quality by significantly reducing its carbon footprint and reliance on conventional fossil fuels. Adopting SAF is a critical step towards a more sustainable and environmentally friendly aviation industry, fostering a cleaner future for both the environment and human health. All these advantages and benefits make it very important to do additional studies and experiments to discover the factors that affect SAF emissions which can be done experimentally on the already available TKT-1 test bench currently fed with CAF. This new TKT-1 configuration will add a very important feature since the original three nozzles are enough to run the engine stably allowing us to feed the fourth nozzle with SAF under any blending ration. A new optical window will be installed (near the SAF nozzle) instead of the igniter plug, allowing us to monitor the flam structure.

References

EASA. (2019). *European aviation environmental report 2019*.

EASA. (2023). *European aviation environmental report 2022*.

Noureldin, M., et al. (2015). *Experimental investigation of performance and exhaust emissions of a gas turbine engine fueled with waste cooking oil*. In 18th IFRF Conference-Flexible and clean fuel conversion to industry, Freising, Germany, pp. 1–3.

Chapter 21
Evaluation of UAS-Based Service Business Models for Road Surface Monitoring

D. Szilágyi, D. Sziroczák, Utku Kale, E. Özbek, M. Ayar, and T. Hikmet Karakoç

Nomenclature

GIS	Geographic information system
UAS	Unmanned aerial system
UAV	Unmanned aerial vehicle

21.1 Introduction

Using drones for data collection is fairly common and a well-established practice nowadays. There are challenges beyond the technological aspects (Sándor, 2019; Fedorko et al., 2018), but the number of applications and the volume of drone traffic have been steadily on the rise. They are deployed in various areas ranging from forests (Tang & Shao, 2015) and sea (Joyce et al., 2018) to urban areas (Kim, 2020). This study focused on their application in road data collection and the potential business models that could be adopted to generate profit. The idea of drone road

D. Szilágyi (✉) · D. Sziroczák · U. Kale
Faculty of Transportation Engineering and Vehicle Engineering, Department of Aeronautics and Naval Architecture, Budapest University of Technology and Economics, Budapest, Hungary
e-mail: dszilagyi@edu.bme.hu; sziroczak.david@kjk.bme.hu

E. Özbek
Unmanned Aerial Vehicle Technology and Operatorship, Eskisehir Technical University, Eskisehir, Türkiye

M. Ayar
Aircraft Airframe and Engine Maintenance, Eskisehir Technical University, Eskisehir, Türkiye

T. H. Karakoç
Aircraft Airframe and Engine Maintenance, Eskisehir Technical University, Eskisehir, Türkiye

Information Technology Research and Application Center, Istanbul Ticaret University, Istanbul, Türkiye

T. H. Karakoç et al. (eds.), *Energy and Sustainable Aviation Fuels Solutions*, Sustainable Aviation, https://doi.org/10.1007/978-3-031-70694-3_21

monitoring has been studied, and the proposed concept's feasibility has been shown (Leonardi, 2018). However, not all locations benefit from this similarly, and the economic viability heavily depends on the location and the adapted business model. This study aims to determine which business models are the best for drone-based road surface assessment and monitoring of the Hungarian and Turkish road networks.

Road quality data is already being collected by operators, mainly for maintenance scheduling purposes; this guarantees that these surveys at least partly can be performed by drones, providing a consistent minimum demand. Drones have several advantages over vehicle-based surveys: they can be completed faster and cheaper, and bad pavement conditions or blockages do not hinder their work. They can also reliably finish a survey because they do not have unpredictable traffic problems.

The operation concept uses fixed-wing hybrid or similarly long-range UAVs with high-quality sensors, at least high-resolution cameras, to scan road sections of interest. The collected data is integrated with a geographic information system (GIS) system and analyzed with image processing methods to provide an output in the shape of a desired quality metric. The gathered insight can be utilized in two main categories:

- Periodic road surface quality assessment: road operators can use the provided metrics to monitor the road surface quality and schedule maintenance.
- Road scan for objects or obstacles: immediately after an event that could cause damage or endanger the road traffic, such as a storm blowing obstacles onto the road, temperature dropping severely causing icing, earthquake damaging the road, floods blocking roads, etc., the drones can quickly make a snapshot of the current conditions that can be relayed to vehicles on the road through displays on the road or a phone application, improving road safety.

The question is how to establish an economically viable service that would provide these services, and it boils down to the question of who would pay for what type of service or information to make this profitable. This is the question that was investigated in this study.

21.2 Analysis

The first step in determining the most suitable business model was to list all the possible activities that could be incorporated into the business model. There were three types of activities: sales type, which involves selling the drone or some part of it; service type, which consists of providing information, managing assets, or renting a drone or a part of the drone; and finally support type, which would supplement selling or service business. The activities are listed in Fig. 21.1.

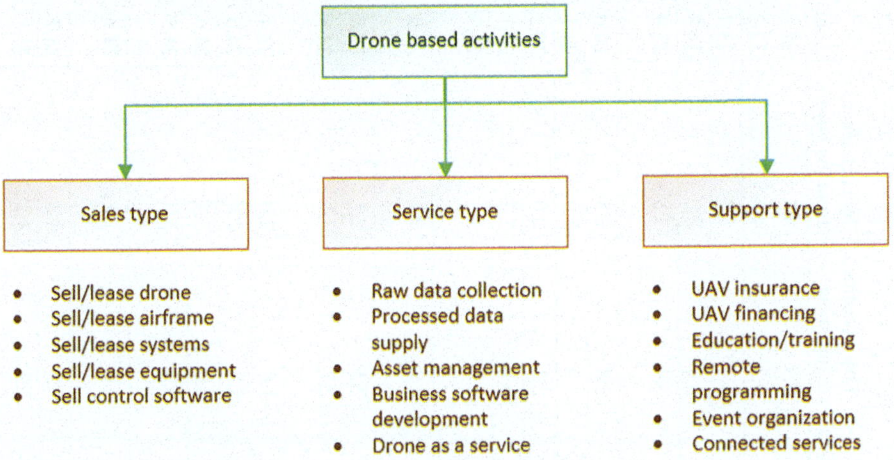

Fig. 21.1 The defined drone-based activities in the three categories

To establish the ranking, a survey was conducted where the business activities were scored by 25 project participants from 1 to 5, with all aspects having equal weight, with the following criteria:

- Expected profit.
- Available market size.
- Lack of competition.
- Difficulty of market entrance.
- Simplicity of technology.
- Expert availability.
- Low investment requirement.
- Independence from the customer.

A score of 1 means very disadvantageous, 5 means very advantageous, and, in the middle, 3 is neutral. The scoring was performed independently, and the participants were asked to briefly summarize their opinions and suggest a model after the scorecard was completed.

21.3 Results

Table 21.1 contains the average scores for all the activities in all aspects, colored with a green scale. The greener the value, the better the score compared to the other moves with green fading to white at the lowest value.

The final score of the activity was chosen to be the average of the eight aspects. Before choosing the best activity, it is worth taking a look at the average scores for each element across all activities. This gives us a rough overview of the market for

Table 21.1 The score matrix of the evaluation with the average scores

Business category	Expected profit	Available market size	Lack of competition	Difficulty of market entrance	Simplicity of technology	Expert availability	Low investment requirement	Independence from the buyers	Sum
Selling drones	3.0	2.5	1.2	2.2	2.5	3.9	2.9	4.4	22.6
Airframe sales	2.5	1.4	1.4	2.3	2.9	3.9	3.3	4.4	22.2
Systems sales	3.5	3.2	1.4	2.1	1.8	4.4	3.4	3.9	23.7
Equipment sales	2.5	3.6	1.9	3.2	3.7	3.9	3.0	3.0	24.8
Control software development	2.8	3.1	1.7	2.7	2.5	3.9	3.4	3.8	23.9
Leasing drones	3.1	2.6	2.5	2.4	3.7	4.4	2.8	2.5	24.0
Airframe lease	2.2	2.0	3.1	2.0	3.4	3.6	2.9	1.9	21.1
Systems lease	2.8	2.3	2.9	2.0	2.2	3.9	2.2	2.0	20.2
Equipment lease	2.3	2.1	3.2	2.2	3.4	4.0	3.2	1.7	22.1
Raw data collection	3.3	3.4	2.6	3.0	3.8	3.7	3.5	1.7	24.9
Processed data supply	4.2	4.0	3.0	2.5	3.7	4.0	2.9	2.7	27.0
Asset management	4.3	3.3	3.6	2.9	2.8	3.8	2.3	2.7	25.8
Business software development	4.3	3.4	1.7	3.0	2.5	4.2	2.7	3.1	24.8
Drone as a service	4.0	3.6	3.4	3.6	2.0	3.0	1.8	2.0	23.4
UAV insurance	3.0	3.8	3.0	2.2	3.3	3.7	2.8	2.8	24.6
UAV financing	3.7	2.9	1.9	2.8	4.1	3.7	2.0	3.4	24.6
Education or training	3.0	2.6	2.3	3.5	3.3	3.8	3.0	3.0	24.5
Remote programming	2.7	2.5	3.7	2.1	3.6	4.2	4.0	1.9	24.7
Event	2.0	2.4	3.0	2.9	3.8	3.9	2.5	3.2	23.7
Connected services	1.9	2.9	2.7	2.7	3.5	3.9	2.9	3.7	24.1
Average	3.1	2.9	2.5	2.6	3.1	3.9	2.9	2.9	

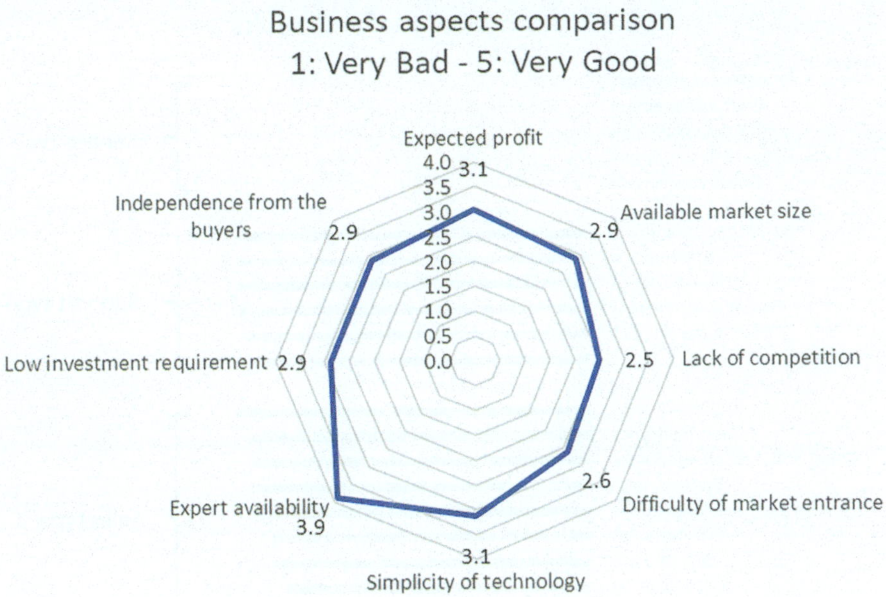

Fig. 21.2 The average of the business aspects across all activities

the investigated operation concept. The average scores for the different business aspects are visualized in Fig. 21.2.

The comparison shows that the strong point of the concept is the availability of required experts, with the expected profit and simple technology also being on the positive side. The two most significant weaknesses are the competition, for example, with ground vehicles, and the difficulty of market entrance. The remaining three aspects, independence from buyers, low investment requirement, and available market size, are close to neutral, very slightly negative.

Looking at the average score of the business activities depicted in Fig. 21.3, we can see that the service-type activities are the best on average. The support category is a close second, with the least advantageous sales type. This is mainly because the selling market of drones or related equipment is a lot more saturated with possibly international competition, whereas providing service locally is less competitive. Activities listed in the support category are less potent because they are usually less profitable and face more competition. However, as a supplementary activity to a primary activity, they could significantly boost the company's revenue. The highest scoring activity was providing processed data to customers: this was the most increased added value activity with the most significant market potential. In the end, both the road operators and the users are looking for readily available extracted information, such as which sections to maintain and where the potentially dangerous sections are, such as damage, ice, or foreign objects where slowing down is recommended, integrated into global positioning system (GPS) navigation, for example. The suggested maintenance actions could be hotfix type, such as

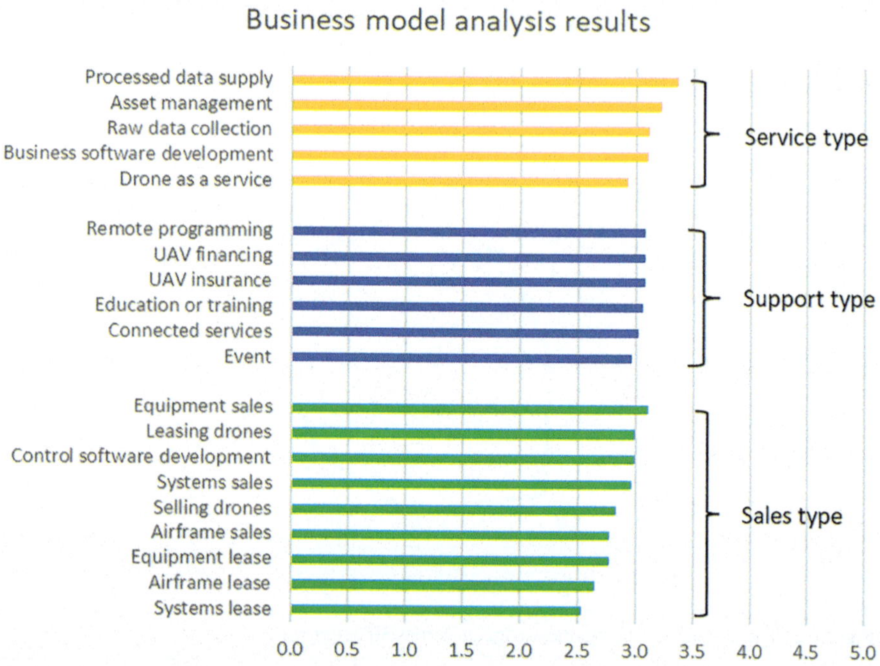

Fig. 21.3 The average score of different activities grouped by type

temporarily filling in a huge pothole or scheduled pavement renovations for a section. In addition, other activities could be considered to supplement the business; good candidates are selling the raw data as well, developing business software, or leasing out equipment that is not used at the moment. According to the majority opinion of the participants, the key to the business's success is creating a cutting-edge data analysis workflow that extracts the information from the collected road surface data.

21.4 Conclusion

Based on the findings of the study, drone-based road surface assessment can be the backbone of an economically viable business. According to the survey, a service-type business is best because selling drones or equipment is a lot more competitive market. The highest scoring activity was providing processed data to customers in the form of indicated potentially dangerous road sections in navigation systems for road users or suggested maintenance requirements with the location of the damage for road operators. The most significant identified risks to the business were the intense competition with ground-based vehicles and the difficulty of the market entrance, meaning that competitors can quickly catch up. The most vital point of the business is the availability of experts, which is critical to developing a cutting-edge data analytics platform that is the basis for the success of this entire business concept.

Acknowledgments Project no. 2020-1.2.4-TÉT-IPARI-2021-00013 has been implemented with the support provided by the Ministry of Culture and Innovation of Hungary from the National Research, Development and Innovation Fund, financed under the 2020-1.2.4-TÉT-IPARI-TR funding scheme.

References

Fedorko, G., Žofčinová, V., & Molnár, V. (2018). Legal aspects concerning use of drones in the conditions of the Slovak Republic within the sphere of intra-logistics. *Periodica Polytechnica Transportation Engineering, 46*(4), 179–184.

Joyce, K. E., Duce, S., Leahy, S. M., Leon, J., & Maier, S. W. (2018). Principles and practice of acquiring drone-based image data in marine environments. *Marine and Freshwater Research, 70*(7), 952–963.

Kim, D. (2020). Pedestrian and bicycle volume data collection using drone technology. *Journal of Urban Technology, 27*(2), 45–60.

Leonardi, G., Barrile, V., Palamara, R., Suraci, F., & Candela, G. (2018). Road degradation survey through images by drone. *Smart Innovation, Systems and Technologies*, 222–228. https://doi.org/10.1007/978-3-319-92102-0_24

Sándor, Z. (2019). Challenges caused by the unmanned aerial vehicle in the air traffic management. *Periodica Polytechnica Transportation Engineering, 47*(2), 96–105.

Tang, L., & Shao, G. (2015). Drone remote sensing for forestry research and practices. *Journal of Forestry Research, 26*, 791–797.

Index

T. H. Karakoç et al. (eds.), *Energy and Sustainable Aviation Fuels Solutions*, Sustainable Aviation, https://doi.org/10.1007/978-3-031-70694-3

The manufacturer's authorised representative in the EU is Springer Nature Customer Service Centre GmbH, Europaplatz 3, 69115 Heidelberg, Germany. If you have any concerns regarding our products, please contact ProductSafety@springernature.com

Printed and bound by CPI Group (UK) Ltd, Croydon, CR0 4YY
30/09/2026
02225604-0007